The Wisdom

of the

TALMUD

A Thousand Years of Jewish Thought

Rabbi Ben Zion Bokser

IAP © 2009

Printed in Scotts Valley, CA – USA.

Bokser, Rabbi Ben Zion.

The Wisdom of the Talmud / Rabbi Ben Zion Bokser – 1st ed.

1. Travel

Book Cover

IAP ©

TABLE OF CONTENTS

INTRODUCTION

The literature of the Talmud represents approximately a thousand years of Jewish thought. Its foundations were laid by the work of Ezra during the middle of the fourth century B.C.E., in the community of the returned exiles from Babylonia, who inaugurated the second Jewish commonwealth in Palestine. Its period of greatest productivity came in the centuries that followed the disastrous Jewish war against Rome in 70 C.E. The Talmud is not an independent literature however. It proceeds instead as a supplement to the Bible. The Bible remained the fundamental source of belief and practice in Judaism, but the Talmud was its authoritative exposition and implementation.

The position of the Talmud in Jewish life has been paramount. It was studied zealously by young and old alike, who found in it the authoritative word concerning the true meaning of Scripture. The lighter side of the Talmud, its parables, its ethical aphorisms, its legendary tales, delighted the common people. The more serious side, the subtle discussions of law, were a welcome outlet for the intellectual interests of the learned.

The Talmud itself became a subject for new commentaries and super-commentaries. Its study commenced in the elementary grades of the Jewish school, and it continued, in ever more subtle techniques of analysis, into the highest grades of the rabbinical academies. The love for the Talmud among the Jewish masses finally created an institution of popular adult education—the voluntary study group that met on Sabbath and holiday afternoons, and weekday evenings, to enable the busy layman to continue his interest in Talmudic literature.

The most crucial element in the discussions of the Talmud is centered in law. For law figured prominently in the Bible, and the Talmud mirrors faithfully the text on which it is based. But the Talmud is not a code. It records varying opinions on law as on life, without always offering decisions as to which was to be deemed authoritative for posterity. The legal discussions of the Talmud are, however, an invaluable source book on Jewish law, for they preserve all the varying trends in the interpretation of Biblical legislation. They likewise preserve a record of new developments in the law by which the Jewish community ordered its life.

The codification of Jewish law was to be a labor of later generations. Utilizing Talmudic discussions as their authority, a group of distinguished scholars, most of them active during the Middle Ages, endeavored to codify the rabbinic law. The most widely used of these works is the Shulhan Aruk by Joseph Karo (1488–1575). This code above all gained popular acceptance, especially in Central and Eastern Europe. But such distinguished teachers of Judaism as R. Moses Iserles, Solomon Luria, Mordecai Jaffe, Samuel Edels and Yom Tob Lippman Heller, did not hesitate to dispute the authority of the Shulhan Aruk. Even as late as the eighteenth century Rabbi Elijah ben Solomon, the Gaon of Wilno, though he had written a valuable commentary on the Shulhan Aruk, did not hesitate on occasion to ignore it and to decide cases on the basis of an original weighing of precedents and circumstances, in the light of the original discussions in the Talmud.

The rise of the Talmud to its dominant role in Jewish life was not without challenge. Its authority was rejected by a group of Babylonian Jews, led by a certain Anan ben David, in the middle of the eighth century. They organized a sect known as the Karaites (from Kara, the study of Scripture), which sought to center Judaism on the sole authority of the Bible. Fierce polemics developed between the Karaites and the Rabbinites, as the defenders of Talmudic authority were called. The Karaites have persisted as a small sect, and several thousand of them still exist in scattered communities in various parts of the world. A Karaite settlement of some five hundred souls has recently been started in the State of Israel, all of them immigrants from Cairo, Egypt. They have been included in Israel's current effort to "ingather the exiles", and they have been recognized as an independent community, free to order life in accordance with its own distinctive interpretations of Judaism.

The non-Jewish world has given the Talmud a mixed reaction. In the Middle Ages, when religious disputations were popular, the Talmud became a frequent subject of controversy. The Talmud was subjected to a variety of criticisms. Because the Talmud had permitted itself to adapt old institutions that they might be more relevant to the needs of a later age, it was charged with the falsification of the Bible. Because the Talmud often speaks in parables, it was disparaged as absurd, as abounding in fairy tales. Because the Talmud reflects a healthy respect for bodily life and speaks with frankness about sex, contrary to the asceticism of the Middle Ages, it was denounced as sensuous and unspiritual.

Perhaps the most serious charge against the Talmud was that it is irreverent toward the beliefs and practices of the Church. The Talmud arose during the epoch when Christianity began its secession from Judaism, and when the Christians were looked upon as dissident Jews. Against that background there must have been extensive controversy between the adherents of traditional Judaism and the advocates of the new doctrine. The Talmud generally avoids polemics; but some echoes of that controversy survived in the Talmud, principally a prayer against sectarianism, the prayer *Velamalshinim*, as it is known in the present Jewish liturgy. This now became a cause of serious charges against Judaism, above all against its revered classic, the Talmud.

The animosity toward the Talmud was often instigated by renegades from Judaism who exhibited the convert's customary zeal by a vilification of the faith they had deserted. The apostate Nicholas Donin laid before Pope Gregory IX the charge that the Talmud was a pernicious and blasphemous work. The Pope responded with an order to seize all copies of the Talmud for an inquiry into their content. In consequence of this agitation twenty-four cartloads of Hebrew books and manuscripts were publicly burned in Paris on June 17, 1242. In the sixteenth century there were six burnings of Talmudic books, in 1553, 1555, 1559, 1566, 1592 and 1599. A Christian censorship of Hebrew books was instituted in 1562. Most editions of the Talmud now extant still carry the censor's assurance that these volumes are free of "offensive" material.

There were voices in the Jewish community that spoke out in defense of the Talmud. Some defended the particular passages of the Talmud which had been attacked. Others addressed themselves to the larger issue involved—they spoke out boldly for religious freedom. One of the most courageous pleas for freedom came from Rabbi Judah Loew of Prague (1512–1609). After analyzing the usual charges against the Talmud in detail and refuting them, he adds: "One ought not reject the words of an opponent. It is preferable to seek them out and study them. Thus shall a person arrive at the ... full truth. Such words should not be suppressed. For every man of valor who wants to wrestle with another and to show his strength is eager that his opponent shall have every opportunity to demonstrate his real powers. But what strength does he show when he forbids his opponent to defend himself and to fight against him? Therefore it is wrong to suppress anyone who wants to speak against religion and to say to him: 'Do not speak thus'. The very converse is true. This itself weakens religion. Suppose the Talmudists did speak against Christian doctrine, expressing publicly what was in their hearts. Is this an evil thing? Not at all. It is possible to reply to them ... The conclusion of the matter is that it would be most unworthy to suppress books in order to silence teachers ..."[1]

The rise of modern anti-Semitism gave fresh impetus to the attacks against the Talmud. The father of the modern calumnies upon the Talmud was the German polemicist, John Andreas Eisenmenger (1654–1704). Eisenmenger offered to suppress his work, *Entdecktes Judentum* (Jewry Unmasked), for a consideration of 30,000 florins, but the Jews refused to be blackmailed into paying this sum. The book has been described as "a collection of scandals" by the *Allgemeine Deutsche Bibliographie*, an official encyclopaedia of German bibliography published by the German Imperial Academy of Science in 1876. "Some passages," the appraisal continues, "are misinterpreted; some distorted; others are insinuations based on one-sided inferences."[2]

The most spectacular campaign against the Talmud was led by August Rohling (1839–1931), a professor of Hebrew Antiquities at the University of Prague. His *Der Talmudjude* (The Talmud Jew) went through 17 editions, reaching a circulation of 200,000 copies in Austria alone. Rohling repeatedly prefaced his slanderous material with the offer of 1,000 Taler "if Judah managed to get a verdict from the German Association of Orientalists that the quotations were fictitious and untrue." The challenge was taken up by Joseph S. Bloch, Rabbi at Florisdorf and later a member of the Austrian Parliament, who offered 3,000 Taler if Rohling could prove that he was able to read a single page of the Talmud chosen at random by Rohling himself. Accusing Rohling of ignorance and perjury, Bloch dared him to bring a libel suit. Because of his professional standing, Rohling could not evade the issue and finally charged Bloch with libel before a Vienna magistrate.

The court was anxious to make a thorough study of the subject and requested the Rector of the University of Vienna, Hofrat Zscholk, and the German Association of Orientalists, to appoint two experts. It conceded to Rohling's request that both these experts be "full-blooded" Christians. Professor Theodor Noeldeke of the University of Strassburg and Professor August Wuensche of Dresden, were selected. From time to time additional experts were called in. After two and a half years, the report was ready. The trial was to start November 18, 1885, but before the hearings began, Rohling, afraid of an open exposure, withdrew all his charges. The court sentenced him to pay the cost of the trial and, disgraced, he was retired from his university post. The entire story of this dramatic encounter is told by Rabbi Bloch in his *Israel and the Nations*.

Another such Talmud "authority" was Aaron Briman, alias Dr. Justus. He was born a Jew and had aspirations for a career as a Jewish scholar. But when he lost face with the Jewish community for deserting his wife and children, he became a Protestant. Subsequently, he became a Catholic and then a Protestant again, and finally tried to return to Judaism. Toward the end of his career he once again joined the Catholic Church. His principal work, published anonymously, was *Der Judenspiegel* (The Mirror of the Jew), a compilation of a hundred laws taken from the Shulhan Aruk and purporting to show the Jewish animosity toward Christians. In a book about the Cabbala, which Briman subsequently wrote under his true name, he said that the whole anti-Semitic literature, including the *Judenspiegel* (his own work!) had been written by stupid and ignorant men. In 1885 he was sentenced by a Vienna court to a long term in prison and expulsion from Austria for forgery of documents. Professor Franz Delitzsch, the famous Protestant theologian, pronounced the *Judenspiegel* "a concoction of damnable lies". Following his expulsion Briman took up medical studies in Paris. These same forgeries of Justus-Briman were later published by another adventurer, Jacob Ecker, who offered them as his own work under the title *The Hundred Laws of the Jewish Catechism*. Czarist Russia made its contribution to this gallery of literary swindlers in the person of the notorious Justin Pranaitis, a Catholic clergyman. His monograph, *The Christian in the Jewish Talmud*, was based on the works of Eisenmenger and Rohling. To create the impression of authenticity he cites many passages in the original Hebrew and Aramaic, but they are all lifted from Eisenmenger, errors and misprints included. By identifying as references to Christians and Christianity such epithets in the Talmud as *am ha-aretz* (literally, a peasant, but more generally, an illiterate person), *akum* (pagan or idol worshipper), *apikoros* (epicurean but applied to heretics generally) and *kuthim* (the Samaritans), he "proves" widespread prejudice on the part of the Talmud toward Christianity.

In spite of his office as a Catholic clergyman, Pranaitis became involved in the course of a checkered career in a series of financial scandals. A picture in a frame which he wanted gilded at the workshop of a certain Avanzo in Petersburg was accidentally damaged; whereupon he tried to extort 3,000 rubles from the owner of the shop on the alleged ground that the picture had been painted by the 17th century artist, Murillo, and that it was part of the collection of Cardinal Gintovt. Both allegations were later proved false. On another occasion he was charged by the board of a local Catholic welfare society in his home parish at Tashkent with misappropriating the sum of 1,500 rubles.

It was in 1912 during the trial of Beiliss on the ritual murder libel that Pranaitis drew world notoriety upon himself by offering his services as an expert for the prosecution. When confronted by the bulls of Popes Innocent IV and Clement XIV which denounced ritual murder charges against Jews as libels and slander and which called upon Christians to desist from the staging of ritual murder trials, Pranaitis denied the genuineness of the documents. Cardinal Merry del Val, the Papal Secretary of State, examined the originals at the Vatican and certified that they were genuine. Beiliss was, of course, acquitted, but the prosecution remunerated the star "expert" with 500 rubles.

Pranaitis died on January 29, 1917. It took more than a month for the Czarist government to issue the permit for the removal of his body from Petersburg to Tashkent. Objections had to be overcome of local officials in Tashkent who were anxious to avoid a public demonstration at his funeral and urged an inconspicuous burial in Petersburg.[3]

Nazi Germany produced a flood of new material vilifying the Talmud. With the pretense of scholarly objectivity which characterized the technique of all Hitler's professors, they produced impressive volumes, but all serving the cause of their master's big lie. Walter Forstat, one of the Nazi "experts" on the Talmud, unblushingly admits in the introduction to his *Die Grundlagen des Talmud* (The Basic Principles of the Talmud, Breslau, 1935), that in writing his book he had really forged a political weapon. "In issuing this work," he writes, "our purpose is purely political … As a political tract it is necessarily one-sided. It therefore deals with Talmudic law only where it may prove helpful in illuminating the attitude of Germany to Jewry."

These diatribes against Talmudic literature produced a reaction, and some of the noblest works in appreciation of the Talmud were written by non-Jews. Johann von Reuchlin and his circle of Christian Hebraists carried on a staunch campaign in the sixteenth century in defense of Hebrew books. The libels of August Rohling were answered by the famous Protestant theologian, Franz Delitzsch, in his work *Was D. Aug. Rohling beschworen hat and beschwoeren will* (What D. Aug. Rohling Has Sworn to and Is Prepared to Swear to, Leipzig, 1883). Among the more recent statements in vindication of the Talmud is the very lucid study by Rev. A. H. Dirksen, "The Talmud and Anti-Semitism", in the January 1939 issue of the *Ecclesiastical Review*, a publication of the Catholic University of America, and the pamphlet, *A Fact About the Jews*, written by the famous Catholic scholar, Joseph N. Moody, and distributed by the Paulist Fathers. A more elaborate study of the Talmud was written by the Polish Catholic scholar, Thadeus Zaderecki. He began his researches in the Talmud under the inspiration of anti-Semitic libels, but what he learned made him into an admirer of this great literature. His work *The Talmud in the Crucible of the Centuries* is a brilliant appreciation of the moral values of Talmudic literature and a refutation of the libels against it, especially those of Rohling and Pranaitis.[4]

Talmudic literature went through a long and varied development. The earliest layer of the Talmud is the *Mishnah*, a product of Palestinian scholarship and written in a clear, lucid Hebrew. The later expository supplement, known as the Gemara, which elucidates the Mishnahic text, was developed during the third, fourth and fifth centuries, when the center of Jewish population was shifting from Palestine to Babylonia. Paralleling the Palestinian *Gemara* there also arose a Babylonian *Gemara*, produced by the newer academies of Babylonia. Both *Gemaras* were written in the Aramaic vernaculars then current in Babylonia and Palestine.

The Talmud has survived in both traditions, the Babylonian and the Palestinian. In their essential procedures and in their underlying doctrines the two Talmuds are similar. But there are some significant differences between them, reflecting in many instances the differing conditions under which the two communities carried on their work. Thus the Palestinians felt themselves in their own country and they regarded the Roman authorities as exercising their power without moral sanction. They therefore ruled the publicans who, as tax collectors, had collaborated with the enemy, as reprobates, and they refused to extend any credence to their testimony in a court of law. The Babylonians, on the other hand, demanded scrupulous adherence to the law of the land, and the tax collector was for them only a civil servant performing his duty, who was as honorable as any other man.

It was in its Babylonian version that the Talmud became so influential a force in Jewish life. The Palestinian Talmud. on the other hand, made but a slight impression on the Jewish world. The circumstances that led to a preference for the Babylonian Talmud have been variously defined. The Babylonian academies functioned under conditions of greater stability and peace and the discussions which emanated from them reveal a greater measure of lucidity and order. What was perhaps more decisive, however, was the fact that the Babylonian Jewish community had overshadowed the Jewish community of Palestine as a center of culture and world influence. As Dr. Louis Ginzberg puts it: "The Babylonians were more successful in establishing the authority of their Talmud in European countries. This success was largely due to the fact that Babylonia, under the rule of the Abbasids, became the center of Arabic culture, and consequently of Jewish culture, since the majority of Jews then lived in Islamic countries."[5]

There is a splendid English translation of the Mishnah, published in 1933, by Reverend Herbert Danby, Canon of Christ Church and Regius Professor of Hebrew at the University of Oxford, England. There is a German translation of the Babylonian Talmud by Lazarus Goldschmidt. The Palestinian Talmud is available in a French translation by M. Schwab. The Soncino Press in London has recently published a new translation of the Babylonian Talmud in English under the very competent editorship of Dr. I. Epstein.

Among the well-known studies of Talmudic literature in English is the *Introduction to the Talmud and Midrash* by Hermann Strack, prominent Protestant theologian and professor at the University of Berlin, and *Talmud and Apocrypha* by the well-known British scholar, the Reverend I. Travers Herford, and also the short but popularly written essay by Arsene Darmesteter, *The Talmud.* An accurate and exhaustive survey of the world outlook of Talmudic Judaism is available in the monumental work, *Judaism in the First Centuries of the Christian Era,* by the eminent Protestant scholar and Professor of Religion at Harvard University, the late George Foot Moore. A brief digest of the contents of the Talmud, with copious quotations, is available in A. Cohen's *Everyman's Talmud.* A short history of Talmudic times is available in the essay by Judah Goldin, *The Period of the Talmud,* in Volume I of *The Jews,* edited by Dr. Louis Finkelstein.

A splendid study of the Palestinian Talmud is offered us in Dr. Louis Ginzberg's recent work, *Pirushim ve-Hidushim b'Yirushalmi* (A Commentary on the Palestinian Talmud, N. Y. 1941). Its three volumes cover the first four chapters of the tractate Berakot, but its extended notes, and an introductory essay in Hebrew and English, offer invaluable insight into the general nature of the Palestinian Talmud and its relationship to the parallel literature which emanated from the Babylonian academies.

The present work attempts to clarify the relationship between the Bible and the Talmud and to trace the forces that continued to inspire the growth of the new literature and that gave it its remarkable popularity in the history of Judaism. It also endeavors to portray the culture of the Talmud through the citation of representative passages. The literature of the Talmud is vast, and the rabbis who composed it often differed in their thoughts. Our characterization of Talmudic culture cannot therefore be exhaustive, but it is hoped that the spirit of the larger work is nevertheless conveyed through these citations.

The material in this work is addressed to the general reader and not primarily to the scholar. The footnotes have therefore been reduced to a minimum. Crucial discussions are documented, but no attempt was made to subject the sources to textual criticism. Historical material which appears in standard works on the subject is drawn on freely, without the citation of the specific sources.

I am grateful to Dr. Louis Finkelstein, President of the Jewish Theological Seminary of America, who has been a constant source of inspiration and encouragement in my studies. Dr. Louis Ginzberg, with whom I studied the Talmud at the Seminary, remained a friend and guide throughout the years, and I am grateful to him for invaluable help. I am indebted also to Rabbis Max Arzt, Michael Higger, Gershon Cohen and a number of other friends who were helpful in the solution of many individual problems in this study. I am also thankful to Mesdames Estelle Horowitt, Sara Jerome, and Bessie Katzman for typing the manuscript, and to Mr. Jesse Fuchs for help in proofreading. Finally I express my indebtedness to my wife for her suggestions, advice and criticism.

THE TALMUD AS LITERATURE

In the library of the world's literary classics, a place of special distinction belongs to the approximately forty volumes which are designated collectively by the name "Talmud". Formidable in size, written in a difficult Aramaic, elusive in many of its discussions, the Talmud has long been an enigma to many. Can the average reader get some idea of what this vast literature is all about, of the men who produced it, of the ideas which inspired them? Can we open a window to permit the modern reader to behold the world of the Talmud, its culture, its way of life?

The authors of the Talmud did not look upon their teaching as an esoteric doctrine, suited only for the few. They sought to reach all men. They sought to reach the common people no less than the professional scholars. The traditional system of Jewish education began the study of the Talmud in the middle grades of the elementary school, and continued it, on ever deeper levels of analysis, to the academies of highest learning. And one of the objectives of that educational system was to cultivate in the student a taste for the Talmud that was to make of its study his avocation throughout life. Its very name—Talmud derives from the Hebrew *lomed* which means study—suggests that it was meant to be a rich and fruitful field of knowledge and research. The prize of the knowledge of the Talmud can be found, but it requires toil; it requires disciplined study. Those who are ready to pursue it with the necessary diligence will find awaiting them a treasury of rare wisdom to reward their labors.

The Talmud came into being as a supplement to Biblical Judaism. It was intended to bridge the gap between the Bible and life. It was a new creation of the Jewish people in response to the facts of a changing world that could no longer be guided by the simple word as enunciated in the Biblical text.

THE BIBLE REQUIRES SUPPLEMENTATION

The Bible continues to command the reverent loyalty of Jews, and of countless others who have learnt to look upon it as the embodiment of their basic religious beliefs and moral ideals. But the very effort to make Biblical religion the basis of human living exposes its insufficiency—at least for those who live in another milieu than the one in which the Bible took form.

The Biblical text often needs clarification. The Bible, for instance, allows the termination of marriage through divorce, without, however, defining the grounds for divorce, the procedure by which it was carried out, or the fate that was to befall the children of the dissolved family. The Bible similarly prohibits work on the seventh day of the week, but it does not define what is meant by work. Are we to infer that writing a letter, marketing or preparing food is to be construed as work? Must the country's armed services go off duty on the Sabbath? Was the priest to halt his Temple duties, and must the rabbi suspend teaching and preaching? Was healing the sick work, and must it be discontinued on the

Sabbath? The answers to these questions must have been common knowledge at the time Biblical law was formulated, but in the course of the centuries that body of unrecorded knowledge was forgotten, and those provisions of the Bible were, therefore, in need of clarification.

The Bible, moreover, could not have anticipated the specific solutions to the many varied problems created by the altered circumstances of a changing Jewish society. The Bible forbids idolatry as a major sin. When the Jews were drawn into the Roman Empire, they were confronted with the civic duty enforced among all Roman subjects of worshipping the emperor. Were they to yield, or incur the consequences of disobedience to Rome? The ritual for initiating a proselyte into Judaism included the offering of a sacrifice at the Temple of Jerusalem, but how was that to be carried out after 70 C.E. when the Temple was destroyed? That was no academic problem, for in the first century, large numbers of pagans continued to join the synagogue throughout the Roman world. Indeed, how was Jewish religious life generally to be conducted after the fall of the Temple, when so much of traditional Jewish piety had centered in the sacrificial cult and the various ceremonies surrounding it?

The changes in Jewish society which necessitated the supplementation of the Bible were not only political; they were also cultural and social. The law that decreed "an eye for an eye, a tooth for a tooth" (Ex. 21:23), as a principle in the punishment of crime, represented justice at the time of its enactment, but it seemed morally reprehensible to sensitive men of a later generation. The prohibition of lending money on interest implies a primitive agricultural economy, where money is generally borrowed for the purchase of necessary tools or consumer goods. It is however, incompatible with the complex requirements of a commercio-industrial economy which depends on investments and banking. Similarly, reflecting the rural society to which it originally addressed itself, the Bible has no explicit provision for a legal instrument validating a commercial transaction. In the absence of a specifically formulated law, local custom or *minhag*, as it was called, often developed to take its place. It is clear, however, that the law could not abdicate to popular improvisation. If the law was to discipline life, it had to be enriched and supplemented with new provisions, to keep pace with a changing world.

Biblical narratives, too, present various theological, historical and linguistic problems that had to be coped with, if people were to master Biblical study, and continue to find in Scripture a source for guidance and authority in their religious life. How, for example, was the Biblical appraisal of the world as "very good" (Genesis 1:31), to be reconciled with the experiences of evil and death, and the repeated disasters to men championing good causes? With whom did God consult when He was quoted as saying, "Let us make man in our image" (Genesis 1:26)? Does that mean that there are several divine powers, or that God is corporeal and endowed with a concrete image? If Moses was responsible for the writing of the Biblical text, how explain the last eight verses of Deuteronomy, which describe his death and extol the quality of his leadership?

The Bible, finally, had not exhausted the creative genius of the Jewish people. The same creative powers which produced those literary masterpieces of the Bible remained alive in

14

the Jewish community and continued to stir men to see new visions and to incarnate them in new creations of culture.

The Bible never became obsolete. Elements of abiding truth shine through all its pronouncements, even when they bear upon them some of the limitations of the people who labored to give the Bible literary form, and of the age in which it arose. But the Bible needed a commentary to close the gap formed by the passing of generations. People who have revered the Bible as the revealed will of God, and have sought to live by its mandate, have therefore generally felt the need of writing commentaries on it. The most imposing of these commentaries is the vast literature of the Talmud.

The characterization of the Talmud as a commentary on the Bible describes the circumstances of its origin, as well as its essential quality. But we must understand the term "commentary" in its broadest sense. It is more than a new exposition of an old document. It is also an original new creation, a means by which the voices of a new age speak out in their discoveries of new truth. That they are willing to speak through a commentary to an older work dramatizes their sense of unbroken continuity with their own past and their acceptance of the Bible as the all-sufficient work for human guidance in the world. The Talmud is thus of value both as literature of Biblical clarification, and as the depository of the newer cultural achievements within the Jewish people during the years in which it took form.

The Talmud is primarily concerned with law, because the Jews looked upon the legislation in the Bible as its most important element. But the Talmud is also rich in many copious discussions in the field of religion, ethics, social institutions, history, folk-lore and science. Thus we define the Talmud as an encyclopedia of Jewish culture; in form, a supplement to the Bible, and in its contents, a summation of a thousand years of intellectual, religious and social achievements of the Jewish people.

THE SANCTIONS FOR BIBLICAL SUPPLEMENTATION

The supplementation of the Bible, in its rich flowering in the literature of the Talmud, was a daring process. It was conceived as a means of fulfilling the law, but it often proceeded in bold new channels. It was in a sense a confession that God's "word" is in some sense not final, and that man must step in to adapt it to the world. Adaptation is akin to change. Dare man "adapt" the word of God? Is it not presumptuous for man to supplement a work through which the Lord hath spoken? The Bible itself seems explicitly to warn against it. Deuteronomy 4:2 speaks out against any tampering with the word of God: "Ye shall not add unto the word which I command you, nor shall ye take aught from it."

The seeming presumption in supplementing the word of God was destined to be an issue on which conservative and progressive schools of thought debated in Judaism. But the spokesmen for supplementation found ample justification for their labor in the hallowed texts of the Bible itself. For the Bible apparently sensed the need of supplementation, and even projected an institution to accomplish it.

An elaborate system of higher and lower courts was established by Moses, while the Israelites were still in the desert, upon the recommendation of his father-in-law, Jethro; and a supreme court was projected as well, to resolve all legal problems which the lower courts could not pass upon. As Deut. 17:8–12 phrases it: "If there arise a matter too hard for thee in judgment ... then shalt thou arise and come unto the priests and the Levites, and unto the judge that shall be in those days; and thou shalt inquire and they shall declare unto thee the sentence of judgment. According to the law which they shall teach thee ... thou shalt do ..." Every branch of doctrine and law, in other words, seemed to be included in the sphere of authority granted to judicial bodies for clarification and adjustment.

The teachers who supplemented the Bible did not limit themselves to interpretations. At times they promulgated new enactments. But even here they did not inaugurate a revolutionary movement in Judaism. A careful analysis of the historical books of the Bible indicated that Jewish authorities in the past had, under certain circumstances, suspended the procedures of Biblical law. Thus the Bible (Deut. 17:6) requires two witnesses to establish the fact of culpable crime; no one was to be found guilty of crime on the basis of his own confession, without corroborating evidence. But Joshua (Joshua 7:24, 25) executed a soldier by the name of Achan when he confessed violating the orders of the commanding general not to loot the city of Jericho after its capture by the Jews. What can explain the conduct of Joshua, except that it was a time of war and martial law superseded the normal judicial procedure? The prophet Elijah, too, seems to have allowed himself to modify traditional law. He offered sacrifices on Mt. Carmel (I Kings 18), when, according to Biblical legislation, all sacrifices were confined to the central sanctuary in Jerusalem. Apparently the opportunity of discrediting the priesthood of Baal seemed to him sufficient reason to modify the traditional procedure of worship. And did not King Solomon suspend the fast on a Day of Atonement in order to hold the dedication of the Temple which he had built in Jerusalem?[1]

It is thus clear that in an emergency traditional law could be suspended for specified or unspecified periods of time. What was justified in the past constituted precedent for the future—if not to abrogate the law, at least to suspend it pending periods of emergency. Even Deut. 4:2, "Ye shall not add unto the word which I command you, nor shall ye take aught from it", was transmuted, through interpretation, into a sanction for the adjustment of tradition. *The word which I command* you was not taken as a reference to the Bible, but to the final formulation of tradition by later authorities. Contemporary authorities in every age, acting in their best judgment, whether to reaffirm or to revise traditional law, represent the ultimate source of guidance in life; and the general public was not to "add" or "take aught" from their decisions.

All these considerations crystallized into the realization that the ultimate authority to guide life cannot be a written text, but the living interpreters of those texts, the custodians of religious leadership in every generation. In the words of the famous Talmudist Rabbi Jannai: "If the Torah had been given in fixed and immutable formulations, it could not have endured. Thus, Moses pleaded with the Lord, 'Master of the Universe, reveal unto me the final truth in each problem of doctrine and law.' To which the Lord replied, 'There are no pre-existent final truths in doctrine or law; the truth is the considered judgment of the majority of authoritative interpreters in every generation.' ..."[2]

The legal powers of a generation's duly authorized interpreters of tradition were looked upon as a function of their office, regardless of their individual merits in piety or scholarship. As a well-known Talmudic homily expounded it: "When the most insignificant person is appointed leader over the community, he is to be treated as the most eminent of persons. It is said, 'Thou shalt come unto the priests, the Levites, and unto the judge that shall be in those days' (Deut. 17:9). Could it possibly enter your mind that a person would go to a judge who was not in his days! The meaning is that you are to be guided by a contemporary authority, whoever he be. As Scripture puts it (Eccles. 7:10), 'Say not, How was it that the former days were better than these.'"[3]

There is a beautiful rabbinic parable which dramatizes man's complete sovereignty in the development of what we may call the supplementary Torah. On one occasion a fierce debate ensued between Rabbi Eliezer and his colleagues on a complicated problem of law. Rabbi Eliezer continued to cite a variety of arguments but his colleagues remained unconvinced. Finally he invoked divine intervention to corroborate his opinion. "'If the law is in accordance with my view,' he exclaimed, 'may this carob tree offer testimony' (by a divine miracle). The carob tree moved a hundred (or, as others related, 400) cubits from its place. They replied to him: 'No proof can be cited from a carob tree.' Thereupon he exclaimed, 'If the law is in accordance with my views, may this stream of water offer testimony.' The stream moved backward from its normal course. They replied to him: 'No proof can be cited from water-channels.' Then he exclaimed, 'If the law is in accordance with my views, may the walls of this Academy offer testimony.' The walls of the Academy began caving in and were already on the point of collapsing when Rabbi Joshua rebuked them, 'If the students of the Torah contend with one another what concern is it of yours?' Out of respect for Rabbi Joshua they did not collapse, but out of respect for Rabbi Eliezer they remained aslope. Finally Rabbi Eliezer pleaded, 'If the law is in accordance with my views, may testimony be offered from the heavens above.' Whereupon a heavenly voice announced, 'What have you against Rabbi Eliezer? The law is in accord with his views.' Rabbi Joshua at once rose to his feet and announced, 'It is not in heaven' (Deut. 30:12). What did he mean by this? Said R. Jeremiah: That the Torah had already been given at Mt. Sinai; we pay no attention to heavenly voices, because Thou hast long since written at Sinai, 'After the majority must one incline' (Ex. 23:2). R. Nathan met Elijah and asked him: What did the Holy one blessed be He do in that hour? He laughed with joy, he replied, saying, 'My sons have defeated me, my sons have defeated me.'" The Torah was given to men and human minds interpreting the Torah in accordance with their best judgments alone define what is or what is not law.[4]

THE METHOD OF MIDRASH HALAKAH

For Judaism the most important portion of the Bible is law and one branch of the supplement to the Bible likewise deals with law. It is in part an attempt to clarify Biblical prescriptions and, through analysis, to deduce general legal principles that would be applicable in new situations. Such Bible analysis was designated by the Hebrew name *midrash*, which may be translated as probing. It was a probing for explanations, provisions and meanings that did not appear on the surface reading of a text but which might be there implicitly, to be discovered through diligent study and research. The midrashic probing of law is technically known as *midrash halakah*, the term *halakah* possibly being derived from a root which means to walk, and therefore, appropriately designating law which charts a way of life.

Frequently this *midrash* defines more precisely the mandate of the Biblical law. Thus the rabbis asked, "What is the meaning of the text, 'The fathers shall not be put to death for the children, neither shall the children be put to death for the fathers'? (Deut. 24:16) If its intention is to teach that fathers should not be put to death for a sin committed by children, and vice versa, behold it is explicitly stated, 'Every man shall be put to death for his own sin!' (*ibid*) The meaning must therefore be: 'Fathers shall not be put to death by the evidence of children', and vice versa."[5]

The *midrash halakah* was, however, equally concerned with discovering in these Biblical provisions, the generalizations that would offer guidance in new situations. This may well be illustrated by the interpretations of Deut. 24:6 and Exodus 21:26, 27. Deuteronomy 24:6 specifies "No man shall take the mill or the upper millstone as pledge; for he taketh a man's life to pledge." This law is specific in its application, but it was clearly designed to protect the poor debtor in his possession of domestic utensils, indispensable in the preparation of food. It was, therefore, generalized to apply to "all tools used in the preparation of food." Similarly the law in Exodus 21:26, 27 provides: "If a man smite the eye of his servant and destroy it, he shall let him go free for his eye's sake. And if he smite out his servant's tooth, he shall let him go free for his tooth's sake." The specifications, eye and tooth, are seen in their common general aspects as vital irreplaceable bodily organs; and the same law is therefore applied to the mutilation of any organ in a slave's body, which is enough to send him to liberty.[6]

There were times when the midrash could not discover Biblical precedents and it became necessary to legislate, to add to or abrogate traditional laws. The post-Biblical festival of Hanukkah, as well as the organization of the synagogue and the ritual of worship surrounding it, are examples of adjustment in traditional law through the process of legislation. So is the decree suspending all religious observances during the Hadrianic persecutions of Judaism (135 C.E.), except the laws forbidding idolatry, murder and adultery. There were many legislative decrees in the field of civil law, too: the extension of poor relief to poor pagans, the provision that only the least desirable parcels of real estate be taken from orphans in payment of debts, and the institution of universal elementary education in the first century before the common era.

The legislative adjustment of law was described as a *takanna*, an enactment, or *gezera*, a decree. These decrees and enactments were promulgated by individuals or corporate bodies that exercised authority at the particular time. Such legislation has been attributed to Moses, Joshua, David, Solomon, Ezra, as well as to the various subsequent heads of the Sanhedrin, which combined both the supreme judicial as well as legislative powers of the Jewish people.

This legislation did not have the status of Biblical amendments. Conceived as a divinely revealed document, the Torah could not be altered by the hands of men. But this legislation was harmonized with the Torah through a similar technique of midrashic probing which discovered in the Torah itself the sanctions for change.

Occasionally the old law was merely circumvented so that, in a technical sense, its mandate remained intact. This is illustrated by Hillel's reform which did away with the cancellation of debts every seventh year, as provided for in Deut. 15:1–3. This law proved a serious barrier to the development of Jewish trade and commerce. People refused to extend credits and loans for fear that their debts would not be repaid before the general cancellation time. Hillel's remedy, called *prosbul,* was the execution of a document which designated the court as the collection agent, and stipulated that the usual law of debt cancellation on the Sabbatical year shall not apply to this particular loan. The court was not included in the provisions of the Biblical law and was, therefore, technically free to carry on collections until the complete liquidation of the debt.[7]

It was similarly through the circumvention that the rabbis reformed the Biblical code of criminal law. There had developed among the rabbis a strong abhorrence of capital punishment. The Bible, of course, recognized a wide variety of crimes for which the death penalty was to be inflicted. Instead of abrogating the Biblical law, the rabbis circumvented it. They limited capital punishment to circumstances which made it practically inoperative. They ruled out all circumstantial evidence, no matter how convincing. They went beyond the Biblical requirement of two eye-witnesses to the crime. The two witnesses were expected to have warned the culprit of the criminality and legal consequences of his projected act; and the criminal was expected to have defied the warning with the assertion that he refuses to be deterred by them![8]

But the midrash, through an ingenious technique of reinterpretation, discovered sanctions for the formal abrogation of old laws as well. Thus the law which decrees that a criminal be punished "an eye for an eye" (Ex. 21:24) was shown to be but an application of the general principle that the punishment must be proportionate to the crime. For, as one rabbi explained, suppose a blind man injured the eye of another person, how shall the law be applied? Clearly there was only one way—compensation; the Biblical injunction is carried out by making the compensation commensurate with the injury. The new legislation universalized this rule of compensation. The Biblical application of the principle was taken as contingent and therefore dispensable, but the principle itself lived on in the new law.

Similarly the institution of the Sabbath was appraised as a means of enhancing human life. Where, therefore, the observance of the Sabbath endangered an individual's existence, it was obviously to be disregarded, for in the words of Rabbi Jonathan ben Joseph, "The Sabbath is delivered in your hand and not you in its hand." The Babylonian teacher Samuel grounded this ruling on the verse: "And he shall live by them" (Lev. 18:5). The commandments of the Torah were to be the means of enhancing life, not for destroying it. This reinterpretation has been traced to the time of the Maccabean revolution against the Syrian-Greeks. The prevailing observance of the Sabbath had endangered the national cause; the enemy simply delayed military operations until the

Sabbath when the Jews would not resist. The modification of the law superseded temporarily the accustomed Sabbath observance, but its essential purpose was vindicated in the national and religious liberation that followed.

This interpretation is of course applicable to all law, for it is the purpose not only of the Sabbath, but of all law, to enhance human life; and all law must therefore be superseded where the broader interests of life demand it. As one rabbi, ingeniously rendering the ambiguous verse in Psalm 119:126 expressed it: "When it is time to do for the sake of the Lord, they voided Thy Torah."[9]

THE METHOD OF MIDRASH HAGGADAH

The supplementary Torah also deals with the non-legal aspects of tradition, with the doctrines and values which are equally an integral part of Judaism. It seeks to clarify various historical, theological, and ethical assertions of the Torah, to rationalize them in the light of current knowledge and prevailing moral ideals, and to derive from them the generalizations that can inspire, guide, and edify life, in the existing conditions under which men lived. The non-legal branch of the supplementary Torah is called *haggadah*, meaning utterances; and the interpretive analysis of the *haggadah* has been designated as *midrash haggadah*.

The following citations will illustrate the nature and function of the *midrash haggadah*. Genesis 12:5 declares: "And Abram took Sarai his wife, and Lot his brother's son, and all their substance that they gathered, and the souls that *they had made* in Haran; and they went forth to go into the land of Canaan." Rabbi Elazar bar Zimra offered the following comment on this: "If all people in the world should attempt to create a single insect they would be unable to breathe the breath of life into it, and here it is said *and the souls that they had made in Haran.* What Scripture really refers to is the proselytes they won to their way of life. And why does Scripture use the term *made* for the winning of proselytes? It is to teach us that whoever draws a pagan close to himself and influences him to become a proselyte, it is as though he had begotten him. And why does not Scripture use the singular *he had made*, instead of the plural, *they had made?* R. Huna suggested that it refers to both Abraham and Sarai. He made proselytes among the men, and she among the women."

The rabbis speculated as to why the Book of Nehemiah was denied an independent place in the Holy Scriptures, but was incorporated into the Book of Ezra (this was the arrangement in the Biblical canon which was accepted at that time). The answer they offered is "Because he thought of his own welfare; as it is said 'Think upon me my God for good' (Neh. 5:19). Another reason is that he spoke disparagingly about his predecessors; as it is said, 'But the former governors that had been before me were chargeable unto the people and had taken of their bread and wine, beside forty shekels of silver.'" The discussion of the authorship of the different books of the Bible also led to the question who composed the last eight verses in Deuteronomy, which describe the death of Moses, and they were ascribed by some to the disciple of Moses, Joshua.[10]

Midrash haggadah is frequently a defense of traditional doctrines against the challenge of contemporary skepticism. On one occasion Rabban Gamaliel was challenged: "You rabbis declare that wherever ten people assemble to worship, the Divine Presence abides amongst them. How many presences of God are there?" Gamaliel called his interrogator's servant and struck him. "Why didst thou allow the sun to enter and heat the home of your master?" "But the sun shines all over the world," the servant protested. Rabban Gamaliel drew the point of the analogy. "If the sun which is only one of the million myriads of God's servants, can be in every part of the world, how much more so can the Divine Presence radiate throughout the universe?"

On another occasion Gamaliel was taunted, "Your God is a thief, because it is written, 'The Lord caused a deep sleep to fall upon Adam and he slept; and He took one of his ribs'" (Gen. 2:21). In this instance Gamaliel's daughter volunteered the answer. "Thieves," she related, "broke into our house during the night and stole our silver goblet but left a golden one behind." "Would that such a thief visited me every day," the skeptic exclaimed. "Was it not a splendid thing then for the first man when a single rib was taken from him and a mate was supplied to him in its stead?" she retorted.

Midrashic probing was similarly utilized in the new formulations of doctrine. This is well illustrated in the famous homily on the theme of human equality. Probing into all the implications of the verse "Ye shall therefore keep My statutes and Mine ordinances, which if a man do he shall live by them" (Lev. 18:5), one teacher asked: "Whence may it be demonstrated that a pagan, when he conforms to the moral law of the Torah, becomes the equal of a High priest in Israel? From the words, 'which if a *man* do he shall live by them', the term man being universal and referring equally to Jew and pagan. Similarly it is said 'This is the law of *mankind*, Lord God' (2 Samuel 7:19, a possible rendition of the original Hebrew)—it is not stated, 'This is the law of priests, Levites and Israelites, but the more inclusive term *the law of mankind*.' In similar manner, too, Scripture does not say, 'Open the gates, that priests, Levites, and Israelites may enter,' but, 'Open the gates that a righteous *goy* keeping faithfulness may enter' (Is. 26:2)—*goy* means a people or nation generally, Jewish or pagan. And again, it does not say, 'This is the gate of the Lord, Priests, Levites and Israelites shall enter into it', but 'the *righteous* shall enter it', which is more universal (Ps. 118:20). Likewise, it does not say, 'Rejoice in the Lord, O ye priests, Levites and Israelites', but, 'Rejoice in the Lord, O ye *righteous*' (Ps. 33:1). And finally it does not say, 'Do good, O Lord, to the priests, Levites and Israelites,' but 'unto the good' (Ps. 125:4), which clearly refers to good men among all nations. It is thus abundantly demonstrated that even a pagan, provided he adheres to the moral discipline of the Torah is the equal of the highest ranking priest in Israel."[11]

THE MIDRASH AS LITERATURE

The earliest literary form we have of the supplementary Torah is the Midrash, which proceeds through a deduction of law or opinion through Biblical interpretation, and it is organized as a running commentary on the books of the Bible. It includes *halakah* and *haggadah* in accordance with the varying contents of the respective books in the Bible. A number of these early *Midrashim* have been preserved to our day. The best known are the Midrash on Exodus, *Mekilta*, the Aramaic for measure, rule or norm; the Midrash on

Leviticus, *Sifra*, abbreviated from *Sifra de-be Rab*, the Book of the School; and the Midrash on Numbers and Deuteronomy, *Sifre*, similarly abbreviated from *Sifre de-be Rab*, the Books of the Schools. All these books were composed in the early part of the second century, by teachers who for the most part remained anonymous but who functioned in the great academy of Torah studies which was established in Jabneh after Jerusalem's fall in 70 C.E.

The following citations illustrate the method of the Midrash and the results in halakah and haggadah achieved by it. When the Israelites on their way out of Egypt found themselves in the difficult position between the pursuing hosts of Pharaoh and the menacing waters of the Red Sea, Moses turned to God in impassioned prayer. But the Lord responded with a sharp rebuke: "Wherefore criest thou unto me? Speak unto the children of Israel, that they go forward" (Exodus 14:15). Rabbi Eliezer elaborates on this: "Thus did God speak unto Moses: 'Moses, my children are in great distress; they are hemmed in by the sea on one side and the pursuing enemy on the other. And yet you stand and indulge in prolonged prayer. Wherefore criest thou unto me? There are occasions when it is proper to prolong and there are occasions necessitating action when prayer is to be abbreviated.'"

Another revealing example is offered us in the comment on Exodus 18:12. This Biblical verse reports: "And Jethro, Moses' father-in-law, took a burnt offering and sacrificed for God; and Aaron came and all the elders of Israel to eat bread with Moses' father-in-law before God." The rabbis wondered why Moses was not mentioned in the episode, and they inferred: "He must have stood by to wait on them and serve them. In doing this he followed the precedent of father Abraham who personally waited on the three angels who came to him in the disguise of itinerant strangers. Similarly, when Rabban Gamaliel arranged a banquet in honor of his fellow scholars, he stood by personally to wait on them and serve them. Some felt reticent, regarding it as improper that they be waited on by the head of the Sanhedrin. But Rabbi Joshua reassured them, 'Let him serve. There is the precedent of a greater man than he who served the three angels who came to him posing as pagan Arabs.'"

Another illustration of the midrashic method may be found in the rabbinic discussion of Exodus 20:18. "'And the people stood afar off; but Moses drew near unto the thick darkness where God was.' What was responsible for this unique distinction accorded Moses? His humility, as it is written (Nu. 12:3) 'and the man Moses was very humble'. The general inference suggested by this verse is that whoever is of a humble spirit will in the end experience the presence of God, as Isaiah (57–16) also suggests, 'I dwell in the high and holy place, with him also that is of a contrite and humble spirit ...' But he that is proud and arrogant renders the land unclean and causes the withdrawal of God's presence, as it is written, 'Every one that is proud in heart is an abomination to the Lord' (Proverbs 16:5), the very phrase used in Deuteronomy 7:26 to describe idolatry."

The following citation is a good example of the midrashic analysis of the Bible for the derivation of law, "'And if a man come presumptuously upon his neighbor, to slay him with guile; thou shalt take him from Mine altar, that he may die' (Exodus 21:14). The verse obviously excluded from the prescribed punishment those who cannot be said to have acted with presumption, such as one who is deaf and dumb, an imbecile or a moron … the physician accidently causing death while working for the patient's recovery or the executioner inflicting death under the order of the court."

The following selection from the Sifre shows the same methodology as the above. "'And it shall come to pass, if ye shall hearken diligently unto My commandments … to love the Lord your God' (Deut. 11:13). One may be tempted to say, 'I will study the Torah so that I become rich, that I may be called master, that I may receive rewards in the world to come'. It is for this reason that the verse emphasizes, *to love the Lord your God;* whatever you do, let it be only with the motivation of true love."[12]

The Midrash was particularly suited for preaching purposes. In following the continuous text of the Bible it enabled the preacher to draw his lesson each week from the Scriptural lesson designated for that particular Sabbath. For the scholar it was also an advantage to see in each instance how a particular law or moral utterance is traced to its Biblical source. From the standpoint of those interested in law, however, the Midrash is an awkward literary form. Relevant legal material is scattered throughout the Biblical books. The presentations are long and involved. The formulation of law is constantly interrupted by haggadah, by historical, theological or homiletical discussions. Nor was the Midrash a convenient depository for enactments and ordinances promulgated without specific reference to Scriptural derivations. There was obviously a need for a work of literary reorganization that would separate halakah from haggadah, that would reduce the law to simple succinct statements, systematically organized along thematic lines, and that would include, also, independent legal traditions which the Talmudic supplementers inherited from the past.

THE MISHNAH

The next product in this process of literary creativity is the Mishnah, a term derived from *shanah*, which means to repeat or study. This Mishnah in the form that we have it today is a product of the scholarly editorship of Rabbi Judah the Prince, and his Palestinian disciples who were active in the 3rd century. But there were other Mishnah collections which paved the way for their labors, going in some instances back to the 1st century. In this final product the language is a clear and lucid Hebrew; the statements are succinct and the principle of organization is subject matter. There are six main sections to the Mishnah which are in turn subdivided into an aggregate of 63 tractates. The tractate is subdivided into chapters and the chapter into individual paragraphs or *Mishniot*.

The six main sections to the Mishnah are called Sedarim, orders, derived from the fact that each section represents an orderly arrangement of the laws on its particular subject. The six Sedarim are *Zeraim, Moed, Nashim, Nezikin, Kodashim, Toharot. Zeraim,* or seeds, deals with agriculture; appended to it is the all important tractate *Berakot* which deals with prayer. *Moed,* festivals, deals with the Sabbath, holidays, fasts and feasts of the Jewish calendar. *Nashim,* which means women, discusses marriage, divorce and other phases of family life. *Nezikin,* injuries, deals with civil and criminal law. *Kodashim,* Holy Things, discusses the sacrificial cult and other details of the Temple service. The last section, *Taharot,* cleanliness, deals with all questions of ritual purity.

The stylistic and methodological character of the Mishnah is well illustrated by the following selections: Mishnah Gittin 9:3, 4 discusses inadequately executed documents of divorce: "Three kinds of bills of divorce are invalid, yet if she married again the offspring is legitimate; one that a man wrote with his own hand but there were no witnesses to it; one to which there were witnesses but which bore no date; and one which bore the date but had one witness only. Lo, these three bills of divorce are invalid, yet if she married again the offspring is legitimate. Rabbi Eliezer says: Even though it was not signed by witnesses yet was delivered before witnesses, it is valid, and she may exact her *Ketubah* from mortgaged property; for the witnesses sign only as a precaution for the general good."

Mishnah Baba Batra 5:1 discusses the transfer of property: "If a man sold a ship, he has sold also the mast, the sail, the anchor, and all the means for steering it; but he has not sold the slaves, the packing-bags, or lading. But if he had said, 'It and all that is in it', all these are sold also. If a man sold a wagon he has not sold the mules; if he sold the mules he has not sold the wagon. If he sold the yoke he has not sold the oxen, and if he sold the oxen he has not sold the yoke. Rabbi Judah says: The price makes it manifest: thus if one said to him, 'Sell me thy yoke for 200 zuz', it is manifest that no yoke costs 200 zuz. But the sages say: The price is no proof."

The procedure in courts of law is described in Mishniot Sanhedrin 1:1 and 3:7. It reads thus: "Cases concerning theft or personal injury are judged by three (judges); claims for full damages or half-damages, two-fold restitution, or fourfold or five-fold restitution, and (claims against) the violator, the seducer and him that hath brought an evil name (must be judged) by three, so says Rabbi Meir. But the other sages say that the latter should be judged by twenty-three, for there may arise therefrom a capital case ...

"When the judges reached their decision they brought in the suitors. The chief among the judges says, 'Thou, such-a-one, art not guilty', or 'Thou, such-a-one, art guilty'. And whence do we know that after one of the judges has gone forth he may not say, 'I declare him not guilty and my fellows declare him guilty; but what may I do, for my fellows outvoted me?' Of such a one it is written, *Thou shalt not go up and down as a talebearer among thy people* (Lev. 19:16); and it also says, *He that goeth about as a talebearer revealeth secrets, but he that is a faithful spirit concealeth the matter*" (Proverbs 11:13).

Mishnah Abot 5:7 offers a precious insight into the kind of character prized by the rabbis: "There are seven marks of the clod and seven of the wise man. The wise man does not speak before one that is greater than he is in wisdom; and he does not break in upon the words of his fellow; and he is not hasty in making answer; he asks what is relevant and makes answer according to the halakah, and he speaks on the first point first and on the last point last, and of what he has heard no tradition he says, 'I have not heard'; and he admits the truth, and the opposite of these are the marks of the clod."

THE TOSEFTA

The Mishnah as finally compiled was a milestone in the history of tradition. It was the summation, the climax of centuries of intellectual labors. It was welcomed particularly by the Jewish community in Babylonia since it offered them religious guidance without necessary recourse to the academies of Palestine. But the process of judicial creativity did not cease with the creation of the Mishnah. Because of its very brevity, the statements in the Mishnah required constant amplification and interpretation. Moreover a great deal of material was omitted altogether, whether because the editors of the Mishnah did not consider it important or because they felt they had already covered it in another form. Such material was technically known as "Baraita", outside, that is, relevant data left outside the Mishnah text. For a scholarly grasp of the full range of tradition it was, however, invaluable.

The shortcomings of the Mishnah must have been recognized in the very generation that produced it; the Tosefta, meaning *supplement*, compiled apparently during the same period that saw the reaction of the Mishnah, frequently offers essential amplifications to the Mishnaic text, as well as, of course, certain independent material. The Tosefta's function as a supplement to the Mishnah is well illustrated by a comparison of Mishnah Shekalim 1:1, and Tosefta Shekalim 1:1. Thus the Mishnah: "On the first day of the month of Adar announcements are made concerning the payment of the annual half-shekel due to the Temple Treasury (Exodus 30: 13 ff). ... And on the fifteenth day of that month the roads are repaired ..." Apparently presupposing the Mishnah, the Tosefta merely amplifies: "On the fifteenth day of that month emissaries of the court attend to the repairing of the roads, which have become damaged in the rainy season." Without the Mishnah we should indeed remain in the dark as to what the Tosefta meant by *on the 15th day of that month*. It is only by drawing on the information of the Mishnah that we may identify *on the 15th day of that month* as referring to the month of Adar.

THE GEMARA

The most important supplement to the Mishnah is the *Gemara*, created after the completion of the Mishnah in the third century. Derived from the Aramaic *gemar* and meaning study or teaching, the Gemara exists in two versions, both in the Aramaic vernaculars current respectively among the Jews of Palestine and Babylonia. For in post-Mishnaic times, the Jewish community in Babylonia had overtaken Palestine as a center of Jewish learning, and the Babylonian schools developed a parallel supplement to the Mishnah, which indeed proved even more influential than the Palestinian supplement. Frequently the same teachers are represented in both Gemaras, for there was a constant interchange of visits among the Palestinian and Babylonian rabbis and the academies in each country were fully informed on the work being done by their sister academies in the other country. Not all tractates of the Mishnah are supplemented by the Gemara—only those that were of interest to the teachers that created the Gemara. The Palestinian Gemara, frequently called Yerushalmi or Jerusalem Gemara, supplements thirty-nine tractates; the Babylonian only thirty-six and a half. In scope, however, the latter is three times as large as the former, the Babylonian Gemara being more elaborate and more copious in its expositions.

In its discussions the Gemara introduces citations from the Tosefta, the various Midrashim, the records of old customs, legislative enactments and ordinances, haggadic discourses and ethical observations. A tendency to digress and interpolate various *obiter dicta* in halakah and haggadah has, in addition, enriched the Gemara with a vast store of anecdotes, parables and folk lore. The Gemara is thus the most comprehensive of all the texts in the supplementary Torah. The Mishnah and Gemara, as an integrated text, taken together comprise the Talmud.

The following Talmudic selections illustrate the style and method of the Gemara and its supplementary relations to the Mishnah. The Mishnah declares: "Seven days before the Day of Atonement the High Priest was removed from his home and confined to the office of the counsellors ... They delivered to him elders from the elders of the court and they read before him (throughout the seven days) from the ritual of the day. They said to him, Sir High Priest, read yourself with your own mouth, perchance you have forgotten or perchance you have never learnt ..."

To this declaration of the Mishnah there now follows a Gemara supplement: "It is understandable that they assume he may have forgotten, but would a High Priest ever be appointed if he had never learnt? Has it not been taught: The Torah describes the High Priest as *the priest that is highest among his brethren* (Lev. 21:10, 14), which means that he must excel his colleagues in vigor, in personality, in wisdom, and in financial independence ... R. Joseph explained: This is no difficulty. The one characterization refers to the High Priests who functioned in the First Temple. The other refers to the corrupt High Priests who held office in the Second Temple. As is illustrated in the report of Rab Assi: A *tarkubful* of *dinars* did Martha, the daughter of Boethus give as a bribe to King Jannai (a general designation in the Talmud for Hasmonean or Herodian rulers) to nominate Joshua b. Gamala as one of the High Priests."

From the same tractate is also drawn the following Mishnah: "A sick person is fed (on the Day of Atonement) at the word of experts, and if no experts are there one feeds at his own wish till he says: 'Enough'". And there follows immediately the vital supplement of the Gemara: "Rabbi Jannai explained: If the patient says, I need food and the physician says that he does not, we hearken to the patient. What is the reason? *The heart knoweth its own bitterness* (Prov. 14:10). But isn't it self-evident? We might have assumed that the physician's knowledge, being more authentic, ought to carry greater weight. If the physician says that he needs food, while the patient says that he does not, we heed the physician. Why? There is always the fear that the patient may be in stupor."[13]

Here is another Mishnah: "If debris falls on someone and it is doubtful whether or not he is there or whether he is alive or dead, one should open the heap of debris to rescue him, even on the Sabbath." The Gemara supplement follows: "One must remove debris to save a life on the Sabbath, and the more zealous one is in doing so the more praiseworthy he is; and one need not seek permission from religious authorities. How so? If one saw a child ... fall into a pit, he breaks loose one segment (of the entrenchment) and pulls it up—the faster the better; and he need not obtain permission from religious authorities. ... If he saw a door closing upon an infant thereby frightening or endangering the infant, he may break it so as to get the child out—the faster the better; and he need not obtain permission from religious authorities. ... One may extinguish or isolate the flames in the case of a fire—the sooner the better; and he need not obtain permission from the religious authorities.

26

"Rabbi Ishmael, Rabbi Akiba and Rabbi Eleazar b. Azariah were once on a journey, with Levi ha-Saddar and Rabbi Ishmael, the son of Rabbi Eleazar following them. This question was asked of them: 'Whence do we know that in the event of danger to human life all laws of the Sabbath are superseded?' Rabbi Ishmael answered and said: '*If a thief be found breaking in*' (Ex. 22:1), it is permissible to kill him in self-defense, though the shedding of blood pollutes the land and causes the divine spirit to depart from Israel. If the defense of life takes precedence over another life—that of the burglar—it certainly takes precedence over the Sabbath. ... Rabbi Simeon b. Menasya said, 'And the children of *Israel shall keep the Sabbath*' (Ex. 31:16). The Torah obviously implied: 'Suspend for his sake one Sabbath, so that he may keep many Sabbaths.' Rab Judah said in the name of Samuel: 'If I had been there, I should have suggested a more convincing explanation. The Torah appraises its rules of life with *He shall live by them* (Lev. 18:5), implying clearly that one must not thwart life because of them.' Raba said: 'The other explanations may be refuted but that of Samuel is irrefutable'".[14]

Our final illustration is taken from the field of civil law. The Mishnah provides: "If a person found something in a shop, it belongs to him; should it have been between the counter and the shopkeeper, it belongs to the latter. If he found it in front of a money-changer, it belongs to him; should it have been between the form (on which the coins are displayed) and the money-changer, it belongs to the latter. If a person purchased fruits from his fellow or the latter sent him fruits, and he found coins among them, they belong to him; but should they have been tied in a bundle, he must advertise."

The Mishnah's discussion of found property evoked the following story from the teachers of the Gemara: "Alexander of Macedon visited King Katzya, who displayed to him an abundance of gold and silver. Alexander said to him, 'I have no need of your gold and silver. My only purpose is to see your customs, how you act and administer justice.' While they were engaged in conversation, a man came before the king with a case against his fellow from whom he had bought a field with its scrap-heap and in it discovered a bundle of coins. The purchaser contended, 'I bought the heap but not the treasure hidden in it' and the vendor asserted, 'I sold the heap and all it contained.' While they were arguing together, the king turned to one of them and asked, 'Have you a son?' 'Yes,' he replied. He asked the other, 'Have you a daughter?' and he answered, 'Yes'. 'Let them marry and give them the treasure', was the king's decision. Alexander began to laugh, and Katzya inquired, 'Why do you laugh? Did I not judge well? Suppose such a case happened with you, how would you have dealt with it?' He replied, 'I would have put them both to death and confiscated the treasure.' 'Do you, then, love gold so much?' said Katzya. He made a feast for him at which he was served with golden cutlets and golden poultry. 'I do not eat gold,' he exclaimed; and the king retorted, 'A curse alight upon you! If you do not eat gold, why do you love it so intensely?' He continued to ask, 'Does the sun shine in your country?' 'Certainly,' was the reply. 'Does rain descend in your country?' 'Of course.' 'Are there small animals in your country?' 'Of course'. 'A curse alight upon you! you only live, then, by the merit of those animals!'"[15]

THE ORAL TORAH AND THE TALMUD

All these literary productions of the academies in Palestine and Babylonia from the close of the Biblical canon to the close of the fifth century comprise the supplementary Torah. This includes the various midrashim; the Tosefta; the Mishnah and the Gemara, or, taken together, the Talmud. It has generally been called the Oral Torah because for centuries it was expounded and transmitted orally. Individual students probably employed notes to aid their memories, but none of these compilations were officially edited until a considerably later date. The Palestinian Talmud came to an end some time in the 5th century as a result of the general decline of the Jewish community in Palestine, marked by the abolition of the office of patriarch, as the head of the Jewish community was called, in 425 C.E. The Babylonian Talmud was concluded toward the end of the same century, for in Babylonia, too, Jewish life was declining, following the persecution of Jews under the Sassanian Kings Yezdegerd II (438–457) and Peroz (459–484). The final edition of the supplementary Torah in the volumes of the Talmud as we have them today, brought to a close one of the most creative epochs in the history of Jewish tradition.

THE FORERUNNERS OF THE TALMUD

THE SOPHERIM

In biblical times the work of supplementing the written Torah was in the hands of priests (Kohanim), Levites and community elders. When the second Jewish commonwealth was founded by the returned Babylonian exiles, that function was taken over by the *sopherim*. The term *sopherim* has generally been translated as *scribes*. As a leader of culture, the *sopher* was usually the one who possessed the then rare skill of writing, and his derivative function was therefore that of scribe. In its original meaning, however, *sopher* was primarily a narrator or teacher, an expounder of a body of tradition; and his work was essentially oral. Thus the verb *saper* which describes the sopher's work designates, in modern as in Biblical Hebrew, the oral activity of narration and instruction.[1]

The pioneer of the sopheric movement was Ezra who came from Babylonia in 459 B.C.E. with the ideal of directing the reorganized Jewish settlement in Palestine toward the principles and institutions of the Torah. The leaders of the new settlement, struggling with the problems of reconstruction, had done little to safeguard the religious interests of the community. They had rebuilt the Temple, but the study and practice of the Torah was widely ignored. And the Temple itself was in a state of moral and material decline. It was this condition that Ezra sought to remedy. At a public assembly described in Nehemiah 9–10, he moved the people to pledge themselves with an oath to "walk in God's law, which was given by Moses the servant of God, and to observe and do all the commandments of the Lord our God and His ordinances, and His statutes." To implement their resolve they also imposed upon themselves certain obligations for which no provision exists in the Pentateuch: a poll tax for the maintenance of the Temple; an arrangement for supplying wood for the altar by the various families in turn during the year; and the institution of priestly supervision over the Levites in their collection of the tithes.

The great task to which the sopherim gave themselves was the popularization of the knowledge and appreciation of the Torah. They instituted the public reading of the Torah not only on Sabbaths and festivals but on those weekdays, Mondays and Thursdays, when villagers gathered in the town markets. They reformed the Hebrew script, introducing the present square type alphabet in place of the old Hebrew alphabet, which is still used by the Samaritans. They enriched the collection of Biblical books by the addition of Ezekiel, Daniel, Esther, the Twelve Minor Prophets, Ezra, Nehemiah and Chronicles. They re-edited the Biblical text bringing it into greater conformity with their developed religious and literary sensibilities. Thus they substituted for the original "but the Lord stood yet before Abraham" the presently accepted reading "but Abraham stood yet before the Lord".[2] They enriched the liturgy with many new compositions and fixed a ritual for the daily and Sabbath synagogue services.

The teachers who supplemented the Torah did not reach their decisions on individual impulse; in every generation there were corporate bodies that deliberated and acted in concert. Such a corporate body functioned during the time of the sopherim, and was known as the Great Assembly. There are no records extant of the sessions of the Great Assembly and we have no significant information about any of its members, except fragmentary echoes in the writing of a later age, principally in the ethical treatise Abot. One of the Assembly's guiding principles is quoted as: "Be deliberate in the interpretation of the law; raise up many disciples; and make a fence about the Torah." One of the last of the men of the Great Assembly, Simon the Just, is quoted as the author of the maxim: "The fabric of civilization depends upon three virtues, the study and practice of the Torah, religious worship, and acts of loving-kindness." Antigonus of Sacho is the only sopher we know of who lived during the early Greek period; and of him only one maxim cited in Abot has been preserved. He advocated serving God without the thought of reward. "Be not like the slaves who serve their masters for the gratuity which they expect. Serve without expecting a gratuity and let reverence for God ever be upon you."[3]

The sopheric movement flourished throughout the years that Persia was the imperialist master of Palestine (563–332 B.C.E.). For apart from the exaction of the tribute, Persia left her colonial provinces complete freedom in determining their inner destiny. Persia crumbled in 332 B.C.E. at the blows of the youthful conqueror of Macedonia, Alexander the Great. Alexander did not live long enough to enjoy the fruits of his military exploits. He died in 322 B.C.E., leaving his empire without a successor. After some bitter fighting, Alexander's leading generals divided the spoils among themselves. Seleucus took Syria, while Ptolemy became King of Egypt. Palestine was claimed both by Seleucus and Ptolemy. Several bitter wars were fought over the issue and the country changed hands a number of times, until 198 B.C.E., when by a decisive feat of arms the Seleucid king added her to his realms. The political uncertainties and the actual dislocations of war must have reacted disastrously upon the social and economic life of the country. But the cultural life remained free and the Torah loyalists continued to practice and propagate the knowledge of the Torah throughout the land.

The free development of the Torah was interrupted during the reign of Antiochus IV who ascended the Seleucid throne in 175 B.C.E. Antiochus had spent his youth as a hostage in Rome. His knowledge of Roman imperial ambitions convinced him that Rome would continue encroaching upon his domains and that war was bound to come between the two empires, Syria and Rome. Indeed, the Roman challenge was presented boldly enough in 168 B.C.E. After successfully invading Egypt, Antiochus was forced to leave his rich spoils by a Roman envoy who threatened an immediate declaration of hostilities. To prepare for the challenge, Antiochus sought to consolidate his far-flung territories by fostering everywhere a common Hellenistic culture.

The Greek "culture" which Antiochus sought to foster was not the Hellenic achievements in philosophy, science and the arts, but the uncritical manners, customs, and superstitions of the Greek populace. He built gymnasia for Greek sports and lavish shrines for the various popular deities. The emperor himself, as the embodiment of the state, was proclaimed as divine, taking on a new title *Epiphanes*, a god made manifest, to be adored everywhere as a symbol of civic loyalty and imperial unity. In 168 B.C.E. the practices of the Jewish religion were proscribed on the pain of death; and the Temple at Jerusalem was converted into a pagan shrine, the Jews being called upon to offer sacrifices to a golden statue of Jupiter as their new divinity.

Antiochus had little difficulty in enacting this policy throughout his empire, but in Palestine he met with resistance. Sopheric activity had popularized the love of Torah throughout the land; and when Antiochus sought forcibly to uproot Judaism, everywhere there were people who preferred martyrdom to a betrayal of their religious traditions. The opposition of the Jewish masses was finally articulated by a priestly family in Modin, headed by a certain Mattathias, who proclaimed active resistance. Mattathias soon died, but the struggle was continued by his five sons, around whom rallied bands of Torah loyalists, waging an active campaign of open as well as guerilla warfare against the successive armies sent by the Syrians. The war was prolonged and bitter, but in 165 B.C.E. the Temple was occupied by the loyalist forces and reconsecrated as a sanctuary of the faith of Israel. Complete independence was not achieved, however, till 142 B.C.E.

A variety of factors cooperated to make for the success of this rebellion. There was the brilliant leadership of the Maccabees, as the sons of Mattathias were called, after their oldest brother, Judas Maccabeus. Dynastic rivalries in Syria kept the empire in a state of turmoil and did not permit the king uninterruptedly to pursue the suppression of the Jewish revolt. And, moreover, the Jews had the assistance of Rome, ever anxious to break rival empires so as to be able more easily to swallow the smaller fragments. The interlude of Jewish independence lasted from 142 B.C.E. to 63 B.C.E., when the country was ruled by Maccabean princes who combined the functions of king and high priest. In 63 B.C.E., taking advantage of dissension in the country over the succession to the Judean throne, the Romans under Pompey marched into Jerusalem and proclaimed Palestine a province of imperial Rome.

THE PHARISEES AND SADDUCEES

The successors of the sopherim who carried on the interpretation and development of the Torah during the Maccabean times, were called *Perushim* or Pharisees. The primary meaning of the root *parosh* from which Pharisees is a derivative is "separate," and some historians have rendered Pharisees as "separatists", men who, because of their excessive piety, tended to separate themselves from the common people. But historically, the Pharisees were the popular spokesmen of the people, and therefore could not have kept themselves aloof from them. *Perushim* may also be related to a secondary meaning of *parosh*, interpret, as it is used in Leviticus 24:12.4 *Perushim* may thus be rendered as expounders or interpreters. Like the *sopherim*, they derived their name from their function, the interpretation of the written Torah.

The Pharisees supplemented the written Torah with new clarifications, with new religious and ethical concepts and new legal formulations. They taught the beliefs in retribution in a life hereafter, the immortality of the soul, the resurrection of the dead, and an extensive angelology. They elaborated the Temple ritual with new ceremonials, like the impressive water libation before the altar on the Succot festival. They ordained that the daily *Tamid* sacrifice in the Temple be purchased not from the funds donated by the wealthy few, but from the shekel collections which were contributed by all Israelites. To counteract a

31

popular superstition that God's physical presence resided in the inner shrine of the Temple, they insisted that the High Priest, in his annual entrance to enact the solemn Day of Atonement ritual, omit the incense whose smoke was to screen him from gazing upon God; he was to prepare that incense after his entry into the shrine. Their conception of God was so exalted that they proscribed the pronunciation of His proper name. They moderated the code of Jewish criminal law. All these and various other measures adopted by them were grounded in the recognition that the written Torah must be supplemented by a continuing new tradition which can apply the written Torah's ultimate purposes to the changing facts of life.

The Great Assembly, as a corporate body, perished with the decline of the sopheric movement, but it was reincarnated during Pharisaic times, in the Sanhedrin, a Greek term meaning a court, council or senate. There were various Sanhedrins throughout the country, charged with different aspects of the interpretation and administration of law. The supreme Sanhedrin, consisting of 71 members and originally holding its sessions in the Hall of Hewn Stone in Jerusalem, exercised judicial, legislative and executive functions. It was headed by co-leaders, a president and chief justice, though the head of the state as High Priest could always preside at any court, including the supreme Sanhedrin.

Records have been preserved describing some of the procedure in this august tribunal. The members of the court were seated in a semicircle so that they could see each other and follow the deliberations more closely, with the president in the center and the others to his right and left in the order of seniority. Secretaries recorded the divergent views that developed in these discussions. All court deliberations were public. To enable them to profit by the proceedings, admittance was granted to advanced students who were seated in three rows, also in the order of seniority, and to general students of the law who were placed behind them. The first students in the order of seniority in the first row generally filled each vacancy as it occurred on the bench, with a general promotion following all along the line. A quorum for the transaction of any business was twenty-three judges; and decisions were reached by a majority vote.

Mishnah Abot 1:1 lists the first five co-leaders of the Sanhedrin as Jose ben Joezer of Zereda and Jose ben Johanan of Jerusalem; Joshua ben Perahya and Nittai the Arbellite; Judah ben Tabbai and Simeon ben Shetah; Shemaya and Abtalyon; and Hillel and Shammai. The same Mishnah cites a number of ethical maxims in the names of each of these teachers but otherwise little is known about them. Jose ben Joezer of Zereda taught: "Make your home a gathering place for the wise. Cling to them steadfastly, and avidly imbibe their words of wisdom." Joshua ben Perahya is the author of the maxim: "Provide yourself with a teacher and acquire for yourself a companion; and judge every person sympathetically." Nittai the Arbellite had as his maxim: "Keep away from a bad neighbor; do not associate with the wicked, and do not despair of retribution when it is slow in coming."5 The official opponents of the Pharisees were the Sadducees, who represented the Jewish upper classes, the lay and priestly aristocracy. The Sadducees fought Pharisaism because of an inherent dislike for Pharisaic ideas. As aristocrats to whom the world had not been unkind, they could not appreciate the drive behind Pharisaic insistence on a future retribution in a world to come. They objected to the introduction of popular ceremonials in the Temple cult or the democratization of the daily Temple sacrifice by purchasing it from a common people's fund rather than from the contributions of individual donors, because they had the aristocrats' natural disdain for the common people. They ascribed to the slave status of an animal because they were no doubt the class of slaveholders. They advocated severity in civil and criminal law, because, unreservedly identified with the *status quo*, they could not treat any one who challenged it, with sympathetic consideration.

Their formal rationale was, however, grounded in a strict adherence to the written Torah. While the Pharisees, sensitive to mass needs and mass problems, advocated a flexible interpretation of tradition, whether of the written Torah or the oral traditions promulgated by past authorities, the Sadducees advocated a strict construction of tradition. In their own way, they too had supplemented the written Torah, as is indicated by their reported possession of a special code of criminal law.6 Pharisaic supplementation was, however, much more radical—a more fundamental departure from the past. The Pharisees, of course, rationalized their departure by integrating them, through midrashic links, with Scriptural precedents or by otherwise finding for them Scriptural sanctions. For the Sadducees, however, the midrash was a spurious invention, unauthorized by precedent and a device for undermining the Torah.

As the contemporary historian Josephus relates it, "The Pharisees have delivered to the people a great many observances by tradition from their fathers, which are not written in the law of Moses; and for that reason the Sadducees reject them, saying that only those observances are obligatory which are in the written word, but that those which are derived from the tradition of our forefathers need not be observed. And concerning these things it is that great disputes and differences have arisen among them; the Sadducees are able to persuade none but the rich, and have no following among the populace, but the Pharisees have the multitude on their side."7

Their strict adherence to the written Torah also explains the origin of the name Sadducees. The widely held theory that the name Sadducee goes back to the high priest Zadok mentioned in the Bible (I Kings 1:32) does not seem plausible. Zadok was High Priest in the reign of King Solomon, and why would a party go that far back for a name by which to identify itself? Why, moreover, would a party dominated by members of the Maccabean dynasty choose the name of the traditional High Priestly family, thereby reflecting on their own legitimacy? Sadducee, the Hebrew *Zduki*, may be taken as a derivative of the Hebrew *Zaddic* which means *righteous*, a construction following the parallel form *Shmuti*, which means a follower of the principles of the Teacher Shammai. The Sadducees called themselves by that name because of their conviction that their platform alone represented loyalty to the Torah, that Pharisaism was unconstitutional, a wicked distortion of the true ideals of Jewish religious and social life. The name Sadducees is used in the Talmud interchangeably with the name Boethusians because the contemporary leaders of the Sadducees were the priests of the famous Boethus family.

The early Maccabees who came to power on the wave of a popular uprising against the Syrian-Greek challenge to the Torah were naturally sensitive to Pharisaic principles. In the course of time, however, the Maccabean state gradually changed into a petty dictatorship waging endless warfare with its neighbors, with all the military, civil, and ecclesiastical power concentrated in the hands of the head of the state. Then an ever widening rift ensued between the government and the Pharisees.

This rift is the theme of a number of stories in Josephus and in the Talmud. According to one story, John Hyrcanus (King from 135–104 B.C.E.) invited the leading Pharisees to a banquet in the course of which he asked them for a frank appraisal of his reign. One of the Pharisees told him that he ought to content himself with civil power and resign the office of High Priest. Challenged, this Pharisee produced a technicality because of which he regarded John as disqualified from the high priestly office; his mother, according to rumor, had been a captive of war before his birth which made his legitimacy doubtful. The rumor was proven false and the king demanded that the Pharisees avenge his insult, but they pronounced upon their bold colleague a light sentence. Infuriated, the king broke with the Pharisees and joined their Sadducean rivals. With minor variations the same story is related in the Talmud, but its moving figure is not John Hyrcanus but Alexander Jannai (King from 103–76 B.C.E.).

According to another story a slave of King Alexander Jannai had committed murder. The Pharisaic Chief Justice, Simeon ben Shetah, ordered the King to appear for a trial, at the same time warning his colleagues on the bench to interpret the law with impartiality and not to be intimidated by the royal defendant. The King appeared, but he defied the usual courtesy due to the court and remained seated during the proceedings. As the witnesses prepared to testify, Simeon ordered the King to stand, in accordance with the prevailing court procedure. But the King replied that he would stand only if the other judges concurred in the request. The other judges were, however, intimidated and remained silent, their cowardice provoking a sharp rebuke from the chief justice.8

The Pharisees were the party in opposition throughout the greater part of King Jannai's reign. Those most conspicuous among them were hunted down and persecuted. Many were forced to flee the country. With the collaboration of the Sadducean nobility, the government pursued the expediencies of state without being hampered by too meticulous a consideration for the idealistic principles of Pharisaism. The Pharisaic Sanhedrin carried little official authority except insofar as the people, recognizing the Pharisees as their spokesmen, voluntarily abided by their interpretations of the law, and frequently forced those interpretations upon the reluctant government officialdom. Thus when Alexander Jannai, in exercising his high priestly functions, once performed the Succot ritual in defiance of Pharisaic teachings, the populace demonstrated, pelting him with the citrons which they had brought with them in celebration of the holiday.

At the end of his reign, Jannai realized the overwhelming popularity of the Pharisees among the masses of the people. Before his death he counselled his queen Salome Alexandra, who was to succeed him, to seek a reconciliation with them. And upon her ascent to the throne, she recalled the Pharisees to power. The Pharisaic leader Simeon ben Shetah, who was also her own brother, became the principal minister of state, and Pharisaic ideals, including a quiescent and peaceful foreign policy, dominated the government.

Upon the death of Alexandra (67 B.C.E.), her elder son Hyrcan II, who had been high priest, ascended the throne, with the consent of the Pharisaic ministers. But the leadership of the army and the nobility rallied around his younger brother, Aristobulus, and plunged the country into civil war. The Pharisees did not want to hold power at the cost of civil war and were prepared to accept Aristobulus. Indeed, some of them, disgusted with the degeneration of the Maccabean dynasty, favored the abolition of the monarchy altogether. The struggle was finally resolved through the intervention of the Roman general Pompey who added Palestine to the Roman Empire.

THE JEWS AND THE ROMANS

The Romans respected Jewish autonomy in all cultural and religious affairs. Roman coins used in Palestine were minted locally so as to eliminate the emperor's image which was offensive to Jewish religious sensibilities. Roman law guarded the sanctity of the Temple, and pagans, including Roman citizens and soldiers, were barred from its precincts on the pain of death. Even the garrison that was stationed in Jerusalem left its standard in Caesaria in order not to offend the Jewish community with their pagan and idolatrous emblems. Jewish courts retained their autonomy and administered the law in accordance with Jewish procedure.

Outside of Palestine, too, the Roman empire treated the Jewish community with sympathetic understanding and respect. Jews were exempted from the civic duty of emperor worship; the sacrifice offered in the emperor's name at the Temple in Jerusalem was accepted by the Romans as a satisfactory equivalent. Jewish synagogues were protected by Roman law against sacrilege. Augustus Caesar even went so far as to direct Roman courts not to cite Jews to its sessions on the Sabbath; and poor Jews were given the option of taking a cash grant of money instead of the normal free grant of oil, so that they might not violate the prevailing Jewish dietary laws.

The Roman respect for Jewish cultural and religious autonomy produced a revival in the study of Torah. After the stabilization of the new regime, Pharisaic activity was resumed. Among the most important teachers who functioned during the period of Roman domination were Shemaya and Abtalyon (60–39 B.C.E.), Hillel and Shammai (20 B.C.E.–20 C.E.), Gamaliel I and his son Simeon, and Johanan ben Zaccai.

Hillel was the most influential of the titular heads of the Sanhedrin. A Babylonian by birth, he was attracted to the schools of Shemaya and Abtalyon and made his way to Palestine. He remained to become the recognized Pharisaic leader of his day. His own school drew students from far and wide; among them was Johanan ben Zaccai who was destined to play a crucial role in Jewish reconstruction after the war with Rome.

Tradition loves to contrast his humility and broadmindedness with the severity and narrowness of his colleague Shammai. This is perhaps best illustrated in the story of the pagan who sought to embrace Judaism but insisted on learning its contents while standing on one foot. Shammai dismissed him angrily, but Hillel made him a convert to Judaism. His summary was the formula, "That which is hateful unto you, do not impose on others."9 This is not, as has generally been explained, a negative formulation of the

Golden Rule in Leviticus 19:18. It is a technique for implementing the Golden Rule, by suggesting a more concrete application of it to the problems of human relations.

Mishnah Abot (1:12–14) cites a number of moral maxims in the name of Hillel: "Be of the disciples of Aaron, loving peace and pursuing peace, loving mankind and drawing them nigh to the Torah ... If I am not for myself, who will be for me? But if I am for myself only, what am I? And if not now when?" Among his most important legal reforms was the institution of the *prosbul*, circumventing the prescribed cancellation of debts at the end of every seven year cycle.10 Another of his important achievements was his formulation of seven rules for the midrashic interpretation of Scripture, subsequently expanded to thirteen by a later teacher, Ishmael. The presidency of the Sanhedrin became hereditary in Hillel's family until the final abolition of the office in 425 C.E.

THE NATIONAL DISASTER AND THE EMERGENCE OF THE RABBI

The Pharisaic period in the development of the Torah came to an end with the Jewish rebellion against Rome in 66 C.E. A variety of factors converged to produce that tragic episode in Jewish history.

There was the enormously heavy taxation, imposed by the Romans, particularly upon the people least able to pay it. Specifically the levies included the *annona*, a tax from crops and other farm produce, delivered in kind; a poll tax on males from 14 years and females from 12 years to 65 years; a market tax on necessities of life like meat and salt; various tolls such as on crossing a bridge or entering a city; forced labor and compulsory requisitions of the farmer's animals. The greatest burden of this taxation clearly fell upon lower classes, particularly the rural population. Indeed beginning with Caesar, the Roman tax was as high as 25% of the total crops in the country. Some of these taxes were collected directly, under the general supervision of native officials from nearby cities. Other levies were farmed out to the *publican* whose rapaciousness made him a byword for sin in Jewish society.

The Pharisees called upon their people to keep aloof from their imperialist masters and to spurn their offers of collaboration. Shemaya, who lived shortly after the Romans became masters of Palestine, counselled his people: "Love work; hate mastery over others; and avoid intimacy with the government." The exponents of Torah during this period denounced the Jewish tax farmer as a reprobate and a robber because he collaborated with the Roman system of extortion and oppression. Deceiving the Roman tax collector, they put on a par with deceiving a pirate, for Rome had no moral right to the country which she had occupied by force. As the Mishnah put it: "Men may vow to murderers, robbers or tax gatherers that what they have is Heave-offering even though it is not Heave-offering; or that they belong to the King's household even though they do not belong to the King's household. ..."11

Rome opened vast markets for enterprising merchants and her fiscal policies encouraged shipping and industry with the result that individual families became fabulously wealthy. But the masses of people suffered want. Discriminatory taxation forced many farmers to abandon the land. Some became laborers on the big estates or moved to the cities where they joined the urban proletariat. Many others turned to cattle raising. This movement must have been large since Jewish authorities, probably fearful of a collapse in the economy of Palestine, a thickly populated country requiring an intensive cultivation of the soil, legislated against the raising of cattle, a move not unlike that taken by the Roman emperors beginning with Vespasian when they were confronted with a similar phenomenon in Rome.

The fate of the urban working people was equally tragic. Slave labor never flourished in Palestine as it did in other parts of the Empire; and the humanitarian legislation of the Bible tended to raise the living standard of the slave and the laborer alike. But the absence of a united labor front made for extremely low wages throughout the ancient world, and Palestine was no exception. The skilled worker was not entirely helpless. Thus the Garmu and Abtimas families, Temple bakers and chemists respectively, were able to win substantial wage increases by striking. Their highly specialized work could not be duplicated by strike-breakers who had been imported from Alexandria; and the Temple authorities were forced to accede to their demands.12 The laborer who did not command such specialized skills was entirely at the mercy of his employer, and his earnings could not have been much above the level of mere subsistence.

The laborer, in addition, suffered from the constant threat of unemployment. Josephus records a pathetic attempt to check unemployment through a public works project. To quote Josephus: "... So when the people saw that the workmen were unemployed, who were above 18,000 and that they, receiving no wages, were in want ..., and while they were unwilling to keep them by their treasuries that were there (in the Temple) deposited ... had a mind to expend these treasures upon them; ... so they persuaded him (King Agrippa) to rebuild the eastern cloisters; ... he (Agrippa) denied the petitioners their request in the matter; but he did not obstruct them when they desired the city might be paved with white stone. ..."13

Above all, there was resentment at what the Romans had done to one of the most sacred offices in Jewish religious life, the High Priesthood. For a time, the administration of Palestine was entrusted to native vassal kings, the most important of whom was Herod the Great, who reigned from 37–4 B.C.E. In 6 C.E. the country was placed under the direct administration of Roman governors or procurators. But both the Jewish puppet kings as well as the Roman procurators manipulated the selection of the High Priest so as to have in that influential position a friend and willing collaborator of government policy. Herod made and unmade seven High Priests in the course of his reign. Valerius Gratus, who served as procurator from 15–26 C.E., made and deposed of five High Priests in succession. To emphasize their control over the office of the High Priesthood, the Romans kept the High Priest's vestments in their custody and released them only on important Temple celebrations. The greatest dignitary of Jewish ecclesiastical life was thus reduced to a tool of a foreign imperialism.

The kind of men who would serve in such capacity were obviously not the spokesmen of a high religious idealism but politicians of low moral character to whom the rewards of power took precedence over their duties to their people and their faith. Many of them did not even know how to perform the Temple ritual and it became customary for a

37

committee of the Sanhedrin to coach the High Priest in the performance of the Day of Atonement service for a full week before each holiday. The Mishnah records that this committee would always depart from its mission in tears.14

The worldliness of these High Priests is well described in a number of Talmudic satires. One popular tale reports that the Temple Court used to cry out at one High Priest, "Depart hence, Issachar of Kefar Barkai, who glorifies himself while desecrating the sacred ritual of divine sacrifices; for he used to wrap his hands with silks and thus perform his sacrificial service." Of another High Priest, Johanan ben Norhai, popular legend relates that "he ate three hundred calves and drank three hundred barrels of wine and ate forty *seah* of young birds as a dessert for his meal!" For him the Temple Court cried out: "Enter Johanan ben Norhai to gorge himself with the foods of the altar."15

One teacher, Abba Jose ben Hanin, lamented thus about the various high priestly families of his day: "Woe is me for the House of Boethus; woe is me for their clubs. Woe is me for the House of Anvas; woe is me for their scheming ... Woe is me because of the House of Kathros; woe is me because of their pens (with which they write evil decrees). Woe is me because of the House of Ishmael ben Phabi; woe is me because of their fists. For they are the High Priests, their sons the tax collectors, their sons-in-law the Temple officers, and their servants beat the people with their staves."16

The high priests oppressed not only the people at large but also the humbler members of their own caste, robbing them often of their due share in the priestly perquisites. As Josephus describes the high priests of his time, "they had the hardness to send their servants into the threshing floors to take away these tithes that were due to the priests, with the result that the poorer sort of priests died for want."17

Stirred by a host of unbearable evils, a revolutionary sentiment was developing in the country. The pioneer of the revolutionary movement was the Galilean peasant leader Judas, who "prevailed with his countrymen to revolt and said they were cowards if they would endure to pay taxes to Rome and after God submit to mortal men as their lords."18 The spearhead of the rebellion was the hard pressed peasantry but they were aided by the masses of the people generally, who suffered with them degradation and exploitation.

It was the behavior of the procurator Florus that started the flames of revolt. He had seized17 talents from the Temple treasury and when, in derision, people made an alms collection for him, he ordered the armed forces to attack the citizenry. Matters were patched up and the people even agreed to extend the customary greetings to an incoming troop of Roman soldiers. But when at the direction of Florus the greetings were not returned, rebellion broke loose. The actual precipitation of the struggle was the work of the lower order of priests. They deposed the reigning High Priest and by lots designated his successor, the rural priest Phanias ben Samuel of the village Aphta. With the Temple in their control, the insurgent priests proclaimed the defiance of Rome by rejecting the special Temple sacrifice which had always been offered in the name of the emperor.

But no sooner did the revolution break when it began degenerating into a civil war. Practically the first act of the revolutionary mob was to sack the palace where the archives were deposited and "burn the contracts belonging to their creditors and thereby dissolve the obligations for paying their debts." The upper classes on the other hand remained sympathetic to Rome and were opposed to the revolution. As Josephus innocently reports it, the "great men ... the high priests and men of power ..." were "desirous of peace because of the possessions they had." Unable to cope with the situation themselves, these men "sent ambassadors to Florus and ... to Agrippa (the neighboring Jewish vassal prince) and they desired of them both that they would come with an army to the city and cut off the sedition before it should be too hard to the subdued."[19]

When in spite of their efforts, the revolutionary cause continued to make headway, they pretended conversion to the revolution but sabotaged it from within. They opposed the centralization of authority in the revolutionary leadership, agitating for moderation and "democracy", and they conspired directly with the Roman officialdom. This is perhaps best illustrated by the conduct of the commander of the revolutionary armies, Joseph ben Mattathias, aristocratic priest of Maccabean lineage. Chosen to be the sword of the revolution, he described the revolutionists as "bandits," "the slaves, the scum and the spurious and the abortive offspring of our nation," and he frankly confesses that he never expected them to win in the struggle. His assignment was to organize the defenses of Galilee, but in his autobiography, he admits that the real purpose of his command was to sabotage the revolution. When he discovered that the principal men in Tiberias opposed the war, he slipped to them the telling hint that he agreed with them but that they should be cautious. "I was well aware myself of the might of the Romans; but on account of those bandits I kept my knowledge to myself. I advised them to do the same, to bide their time and not to be intolerant of my command."[20] Josephus actually instigated the escape of a spy sent by the Tiberians to Agrippa, inviting him to come and save their city from the revolution. And at the first opportunity he abandoned all pretense and surrendered to the Romans.

For his treason he was rewarded with high honors: Roman citizenship, a pension, and the right to add the name of the imperial family to his own. Joseph ben Mattathias, the Hasmonean priest, became Josephus Flavius, the favorite of the Romans.

The revolutionaries were not without their calculations when they challenged Rome. When the revolt broke, the Roman empire was shaking with inner unrest. The "affairs of the Romans were ... in great disorder ... the affairs in the East were exceedingly tumultuous ... The Gauls also ... were in motion ... and the Celts were not quiet ..."[21] And the rebels had reason to expect support from the far-flung Jewish communities outside of Palestine. The revolution failed largely because of an old truism in revolutionary history, that no colonial people can hope successfully to wage two revolutions at the same time, a national revolution against its imperialist masters, and a social revolution against its own ruling class. Caught between the cross-fire of the Roman legions without, and the pro-Roman forces of betrayal and counter-revolution within, the doom of the revolution was of course sealed. After a valiant and costly struggle, Palestine was once again a Roman province, with the claws of the imperialist masters more tightly drawn; and the Jewish people resigned themselves to suffer the oppression of an old tyranny and the new ignominy of defeat.

The central fact in the tragic legacy of war and revolution was that the Temple at Jerusalem had been burnt to the ground. For the Temple had functioned as a supreme national shrine in Judaism; its cult of sacrifices was regarded as the principal formula of Jewish worship, and throughout its history it had served as an invaluable symbol of unity and solidarity throughout the Jewish world. The Jews faced a difficult task to reorganize their lives without the resources of the Temple and the hierarchy of institutions that had developed around it.

Jews had once before been called upon to reorganize their religious life without the Temple and they had done so with success. In 586 B.C.E. the armies of Babylon had destroyed the independence of Judah and burnt her national shrine, the Jerusalem Temple. The sobering realizations that followed this national disaster placed the leadership of the Babylonian exiles in the hands of prophetic teachers rather than the princes or the priests. And under their inspiration, the reorganized religious life of the community moved in new directions. The study of Torah and the practices of a personal religious life became central. The synagogue with a ritual synthesizing study and worship, began its long and eventful development.

The reorganization after the war against Rome followed a similar course. The policy of rebellion against Rome had failed miserably. The lay and priestly aristocracy were scattered and discredited. They had been the main targets of the revolutionary terror and their most influential members perished in the civil war that accompanied the revolution. And even when the flames of war and revolution ebbed, the bitterness lingered on, sustained by a vivid record of upper class betrayal perpetrated in the darkest hour of the nation. The Torah alone was left as a rallying point of Judaism and as a possible instrument of post-war reconstruction.

The leader in this movement of reconstruction was Johanan ben Zaccai. He was ideally suited for his task. He had studied under Hillel and the venerable master had proclaimed him "the father of wisdom" and "the father of coming generations". He was chief justice of the Sanhedrin before the fall of the Temple and independently conducted a school in Jerusalem. The popularity of his lectures forced him frequently to speak outdoors where larger gatherings could be accommodated. His profound scholarship was complemented by an equally profound love for human beings. He was ever the first to offer greetings to any passerby in street and market place, and his interest extended to Jew and pagan alike. The greatest of all virtues, he taught his students, was a kind heart which reaches out with sympathy to fellow-humans. The study of Torah was the *summum bonum* in life, but the student of Torah was not to keep himself aloof from the common people. "If thou hast learnt much Torah, ascribe no special merit to thyself; for that is the true function of thy being." On one occasion he deliberately embarrassed himself by pretending he had forgotten an important principle of law, so as to demonstrate to his students a lesson in human fallibility. At the same time, he was endowed with a shrewd practicability, acquired no doubt in the first forty years of his life when he pursued a career in commerce. When his students once asked him for a blessing he offered the prayer: "May ye revere the Lord as ye revere men." "Is that all?" they wondered. "Would that ye revered Him at least in that measure" was his reply, "for see how a person, proceeding to commit an immoral deed, will always say to himself, 'but no man must see me'." Johanan fought hard for Pharisaic control of the Temple, but the cult of sacrifices was not indispensable for his conception of Jewish religious life. In the synagogue and the ritual of personal observance, he found adequate resources for the cultivation of the religious life. For, quoting Hosea (6:6), he explained, the Lord "desired loving-kindness and not sacrifices."[22]

When the challenge of rebellion was ultimately presented, Johanan and his followers counseled submission. They realized that rebellion against Rome would lead to a sanguinary war with untold devastation and tragedy. They believed, moreover, that they could achieve an even more fundamental liberation through other methods. For in the ethical implications of their monotheism, they saw the organic wholeness of the human race.

They consequently hoped not for a national revolution purposing to liberate their people from a foreign government but for a moral revolution to liberate all mankind from superstition, idolatry and falsehood. The ideals of this moral revolution were for them embodied in the Torah and they consequently sought to teach the Torah to natives and pagans alike. Many pagans, from the highest as well as the lowest strata of Roman society, responded to the propaganda on behalf of Judaism and joined the synagogue. Many more, while not officially embracing Judaism, renounced idolatry and became the fellow-travellers of the synagogue, ordering their lives by the Torah's ideals of personal and social morality. It was a slow process but, as Johanan and his followers saw it, it was the only fundamental way of dealing with the problem. And as long as Rome did not interfere with their missionizing propaganda, they were confident that before long the truth would prove mightier than the mightiest legion of Rome.

Johanan left Jerusalem before it was taken by the Romans. He was smuggled out of the city in a coffin by his two most trusted disciples, after feigning illness and death. Thereupon he made his way to the Roman commander and surrendered. Recognizing his influential position in Jewish society, the Romans granted him his freedom and permitted him to reestablish his academy in Jabneh.

With his loyal disciples by his side, he waited breathlessly for the outcome of the struggle around Jerusalem. When the news reached him that the Temple had fallen, he proceeded to act. He proclaimed Jabneh as the new center of Judaism and, with his own disciples and others who joined him subsequently, he recreated his academy and reorganized the Sanhedrin. Study, prayer, the Sabbath and holidays, the cultivation of the spiritual and ethical life, were declared more than adequate substitutions for the cult of Temple sacrifice. And Jewish law, to be promulgated and interpreted at the new Sanhedrin, was to continue to give direction and unity to Jewish life throughout the world.

Until another scion of the Hillel family, Gamaliel II, became available for the office, Johanan bore the title *rabban* by which the titular head of the Sanhedrin was designated since the elder Gamaliel (20–50 C.E.). But to invest his disciples with the new authority of their office he ordained them with the new title, *rabbi*, master. The *rabban* was recognized by the Roman government as the official head of the Jewish community; and in cooperation with the rabbis he directed the study, adaptation and application of the Torah to the new needs of life.23 All the literature of the supplementary Torah, in the form in which it has come down to our own day, including the copious literature of the Talmud, is the work of the rabbis, who became the undisputed leaders in post-war Judaism.

THE TALMUD IN ITS HISTORICAL SETTING

THE AFTERMATH OF WAR

The fall of the Temple had left a void in Jewish religious life. Gladly, the Jews would have labored at its reconstruction, but that was banned by the Romans. Jerusalem which was a symbol of all that was glorious in Jewish history was an armed camp, where Jews were forbidden to enter. They could come as tolerated pilgrims to visit ruined shrines and shed tears over their departed national glory, but they could no longer make their homes there. The half shekel which the Jews had always contributed to the upkeep of the Temple was now collected by the Romans as the *fiscus judaicus*, a special tax upon the vanquished people to be devoted to the maintenance of Roman pagan shrines. There were large scale confiscations of Jewish property, particularly land, which Jews could now occupy only as tenant farmers; and Rome added humiliations to injury by erecting an arch of triumph to Titus and issuing special coins to commemorate the Jewish disaster. "Judaea capta", "Judaea devicta", "captured, vanquished Judea", these coins proclaimed, and they carried, as an illustration of the slogan, the image of a decrepit, broken woman, bowing before her proud conqueror.

These conditions distilled a great spiritual depression in Jewry. Asceticism became widespread. There were those who shunned the use of meat and wine because these had at one time been offered on the sacrificial altar which was now in ruins. Large numbers refused to raise families and beget children, apprehensive of life's uncertainties in a cruel world. And the seemingly unchallenged march of brute power undermined for many the faith in their people's way of life, their beliefs in divine providence, the election of Israel and the supreme worth of the Torah. "If there is a God Who cares for justice, why does He allow all this wrong to go unchallenged in the world?" came the constant cry of questioning multitudes.

RECONSTRUCTION AT JABNEH

The basis of Jewish rehabilitation in this all pervasive crisis had been laid by Johanan ben Zaccai. He shifted the center of Judaism from Jerusalem to Jabneh and launched the new Sanhedrin and the rabbinic movement, rescuing the most important element in the life of a people, centralized direction and authoritative leadership. The nominal head of the Jewish community had been the High Priest whose office perished with the Temple, but he was not indispensable; and the Nasi, the head of the Sanhedrin, stepped forth to replace him. The nasi's office lacked the glamour of the High Priesthood, but he more than made up for it by his piety and scholarship and by his devotion to Pharisaic principles. And he had one more important virtue to recommend him: he was a direct descendant of the famous sage Hillel, and, on his mother's side, of Judah's royal family, going back to King David. Johanan remained the acting head of the Sanhedrin until arrangements could be made for its legitimate occupant to succeed him. Gamaliel II, the legitimate heir of the *nasi*, finally received the recognition of the Roman officials in Syria, and, with the collaboration of a grimly determined but hopeful group of rabbis, including Eliezer ben Hyrcanus, Joshua ben Hananiah and Akiba ben Joseph, he inaugurated a new and colorful chapter in the history of Judaism.

The new *nasi* was well suited for his position. He was a man of independent means, having inherited his family estates in land and slaves, which enabled him to devote himself freely to scholarship and communal work. He was educated in the traditional culture of his people as well as in the worldly knowledge of his day. He was a fine mathematician and astronomer and he had a good command of the Greek language. The Talmud records many anecdotes illustrating his kind and sympathetic character. The joy of having his colleagues as guests in his home was unbounded; and he insisted on taking the place of his servants in waiting on them. He was touchingly devoted to his slave Tebi. Members of his household were trained to call the slave "father" and the slave's wife, "mother". And when Tebi died Gamaliel sat in mourning as for a departed member of the family. "Tebi was not like other slaves," he explained; "he was a worthy man." "Let this be a token unto thee," he once exclaimed, "so long as thou art compassionate, God will show thee mercy; but if thou hast no compassion, God will show thee no mercy."[1]

We do not know the date of his death. Before his passing he left a will which was to convert even his burial into an important lesson for his people. It had been customary to bury the dead in lavish outfits and funeral costs weighted heavily on poor families. Gamaliel therefore ordained that he be buried in simple linen shrouds in the hope that his example would be imitated by others. Thus began a tradition which has endured in Jewry to this very day.

The achievements of the rabbis at Jabneh were varied and far-reaching. To give expression to the universal gloom over the national disaster, they ordained formal rites of grief and remembrance. People were to leave patches of unpainted wall space in their homes; they were to omit some dish from their customary meals; women were to reduce their use of cosmetics and jewelry. This was to remind them that without the Temple their lives were incomplete and that they must ever strive for its restoration.

But there was to be no despair. The rabbis denounced the growing asceticism as inconsistent with the national interest. To abstain from procreating, Rabbi Eliezer ruled comparable to the shedding of human blood. Rabbi Joshua argued with those who were avoiding meat and wine in mourning for the destruction of the altar, by explaining that to be consistent they would have to renounce fruit, bread, and water as well, since they were also used upon the altar.[2]

The destruction of the Temple was a tragic blow to Judaism, but it was not to interfere with an active religious life. Pending the Temple's rebirth, the rabbis proceeded to displace the sacrificial cult with new disciplines for communing with God. They promulgated the famous Eighteen Benedictions as the nucleus of a formalized prayer service which was to be recited thrice daily in private as well as congregational devotions. Some of these benedictions were old and had been recited in the Temple as well as in many synagogues that flourished side by side with it. But they were now re-edited so as to include references to the hoped for resurgence of Jewish freedom and the restoration of the Temple. These prayers, moreover, were now to be recited by every individual worshipper and not alone by the public reader who led in the service. The initiation of a proselyte into Judaism was reorganized, omitting the customary sacrificial offering. The Haggadah, a ritual of narration and dramatic re-enactment of the Exodus, was developed to take the place of the solemn Passover rites in the Temple at Jerusalem.

The rabbis were equally active in counteracting schismatic and heretical tendencies which were making their appeal among the people. They induced Aquila, the Greek proselyte from Pontus, to undertake a new and more literal translation of the Bible which would bring the Greek text into closer harmony with Jewish tradition. The current Greek translation, known as the Septuagint, was too free and inaccurate, making it frequently an easy weapon for Christian and other sectarian propaganda. And they introduced into the religious service a special prayer in denunciation of heresy and heretics. Prayers of denunciation were repugnant to the rabbis who taught the virtues of universal love. To make sure that this prayer would not be inspired by hate or bitterness toward other men, that it would be directed against error rather than against the erring, they entrusted its composition to the saintliest and humblest of their members, Samuel the Modest.

By fearless and searching self-criticism they met the challenge of those who had lost faith in the moral order. The disaster was not an indication of a morally lawless universe, but on the contrary, of the workings of a moral law which cannot be evaded with impunity. Like the prophets of old, they blamed their people's tragedies upon their own mistakes and failures. Jerusalem was destroyed because men hated one another, because her people were not united in the national crisis, because they permitted grave injustices to prevail in their midst.[3] Rome, like Assyria and Babylonia of old, was only the rod of God's indignation, smiting and healing a sinful people. And the disaster itself pointed to the way of redemption. It was for them to repent. to purge themselves of their imperfections, to rebuild their lives on more wholesome foundations, and, in due time, they would be restored to freedom.

Perhaps the most important achievement of the rabbis was the creation of an authoritative Jewish law. The supplementation of the Torah had proceeded ever since the days of the sopherim and a variety of men had contributed to it. The inevitable differences in social and ideological orientation prevailing among men had naturally led to differences in Torah supplementation. But now how was the Torah to guide life if its official interpreters could not agree?

The Pharisees had solved this problem by developing a fine tolerance. All views that developed in the course of their deliberations were regarded as equally sincere attempts to understand and apply the ideals of the Torah to the necessities of life. Men were therefore advised to exercise their own discretion and follow the particular school of thought that best expressed their own conception of right and wrong. "Although," the Mishnah relates[4], "one group permitted marriages which the other prohibited, and declared pure what the other considered impure, they freely intermarried and did not scruple to use each other's food." To signalize this tolerance, the leadership of the

Sanhedrin was divided between the representatives of the majority and minority. The spokesman of the majority became the *nasi*, President, while the leader of the minority group became the *ab bet din*, the chief justice.

After the destruction of the Temple when the Torah was the only surviving institution that could unify Jewish life, the old arrangement was changed. The new Sanhedrin at Jabneh repudiated the old formula of tolerance, except in the field of doctrine. On questions of theology and ethics, individuals remained essentially free to formulate their own doctrines in accordance with the dictates of their own conscience. In the field of action, however, the minority was now to give way to the majority whose views alone were to be promulgated as authoritative law.

The new formula was first applied to the disputes between the School of Shammai and the School of Hillel. By a majority vote, the rabbis, deliberating at the new Sanhedrin in Jabneh, repudiated the Shammaites and declared the views of the Hillelites alone authoritative.

These reforms were not achieved without struggle. One such struggle developed between Rabbi Eliezer and his colleagues. Rabbi Eliezer was one of the pioneers in post-war reconstruction. Together with his colleague Joshua ben Hananiah, he had helped smuggle Johanan ben Zaccai out of the besieged city of Jerusalem and had participated in the organization of the new Sanhedrin. But as the deliberations at Jabneh proceeded, a serious cleavage developed between him and his colleagues. In the general unfolding of his ideology, Rabbi Eliezer followed consistently the general point of view of the Shammaite system of Torah interpretation which had been rejected by a decisive vote of the great majority of rabbis.

The hostility which had been gathering for some time finally culminated in an open break during a discussion about the so-called "Akhnai" stove. According to Biblical law (Lev. 11:33), earthenware, pots and ovens, which had become unclean, for example, through contact with a dead body, were to be broken. The "Akhnai" stove had become exposed to uncleanliness, but the owner cut it into tiles which were separated from each other by sand and externally plastered over with a layer of cement. This was a loose arrangement, but it could still be used as a stove. At the same time, as a "broken" vessel, it would no longer be susceptible to uncleanliness. Rabbi Eliezer's colleagues objected to the arrangement. What was important to them was not so much the objective fact that the stove was "broken", as the manifest intention of the owner who continued to use it. The owner's intention made it again into a "whole" vessel, and its impurity, therefore, persisted. Rabbi Eliezer, who, like the Shammaites, was generally more concerned with objective facts rather than with the intentions behind them, regarded the stove as actually broken, and, therefore, no longer subject to laws dealing with whole vessels. The controversy that raged over this question was prolonged and bitter. Finally the matter was put to a vote, and Rabbi Eliezer was dramatically defeated by a great majority. But Rabbi Eliezer refused to yield. He counselled his followers to defy the majority, and in his judicial decisions continued to formulate the law in accordance with his own views. Behind this impasse stood not only a difference in attitude toward the Akhnai stove, but a challenge to the concept of a disciplined Jewish life.

To break the impasse, the rabbis finally responded with excommunication. In demonstration, they held a public burning of certain types of food which they had pronounced impure, but which he, in defiance of their opinions, persisted in considering pure. Rabbi Akiba, his own disciple, carried the news of the decision to him. Seated in mourning dress, at some distance from him, Akiba spoke: "My master, it appears to me that thy colleagues keep aloof from thee." Rabbi Eliezer understood the message, but remained unyielding to the end.[5]

Rabbi Eliezer felt his isolation most keenly. The terms of his excommunication apparently left him free to continue teaching in his school at Lydda, but he realized that the centre of Jewish learning and authority was at Jabneh. From his pupils, who occasionally attended the sessions at Jabneh, he sought to learn what went on there, but such conversations would only pain him, reminding him that he was an outcast. Once, when he was told that the council at Jabneh had deliberated on a question concerning which he felt himself qualified to speak authoritatively, he actually shed tears, and although the decision of the scholars was in accordance with his own opinion, he dispatched to Jabneh a message of acquiescence. Moved, no doubt, by his own experience, he warned his disciples: "Be as careful about the respect due to your colleagues as about the respect due to yourselves— and do not permit yourselves to become easily provoked to anger"; "Warm yourselves before the hearths of scholars, but see that you are not burnt, for when they bite, it is the bite of a fox, and when they sting, it is the sting of a scorpion."[6]

Rabbi Eliezer was not reconciled with his colleagues until his dying moments. Rabbi Joshua, Rabbi Akiba, and a number of other scholars, hearing of his illness, had come to pay him a visit. They could not draw close—he was still under the ban—and they, therefore, stood at some distance. But he recognized them. "Why have you come?" he demanded summarily. "To study Torah," they replied. "And why have you not come until now?" he continued. Embarrassed, they apologized that they had been busy. Rabbi Eliezer recalled the days when he was still the great teacher in Israel, and looked back upon the time when Rabbi Akiba was still his devoted disciple. To erase the pain induced by these recollections, the visiting scholars drew him into legal discussion. In the midst of it he expired. Forgotten now were all the dissensions; only Rabbi Eliezer's great sincerity, his profound learning and his piety remained. Overwhelmed with grief, Rabbi Joshua arose and formally dissolved the sentence of excommunication that had been between them. Rabbi Akiba applied the verse spoken by Elisha at the passing of Elijah (II Kings 2:12): "My father, my father, the chariots of Israel and the horsemen thereof." A great but turbulent personality had passed from Israel.[7]

There were similar rifts between Rabbi Joshua and the patriarch Rabban Gamaliel II. Exercising his prerogatives as *nasi* to arrange the calendar, Rabban Gamaliel announced the date of the New Year Day. A number of scholars, including Rabbi Joshua who was chief justice of the court, made calculations of their own which led them to different conclusions and they recommended that the date be changed. Rabban Gamaliel, however, refused, regarding the matter as closed. Whereupon Rabbi Joshua proceeded to plan celebrating the Holidays not on the date officially designated, but on the date supported by his own calculations.

Rabban Gamaliel saw the threat of a schism and, to maintain the authority of the court, he ordered his associate "to appear before me with thy cane and thy purse, on the day which is the Day of Atonement according to thy reckoning." Rabbi Joshua was in a dilemma and he came to consult some of his colleagues and friends. Rabbi Akiba advised him to obey, for the authority of the court must be upheld even if its decision was based on technically inadequate testimony. He cited the text describing the jurisdiction of the court in the determination of the calendar. "These are the appointed seasons of the Lord, even holy convocations which ye shall proclaim in their appointed seasons" (Lev. 23:4); their holiness is not inherent, but derived from the proclamation of *the court.* Another colleague, Rabbi Dosa, likewise recommended compliance. "If we are to review the decisions made by Rabban Gamaliel's court," he explained, "we might as well reconsider every decision which was promulgated from the days of Moses to our own." With a heavy heart Rabbi Joshua finally obeyed. The patriarch was overjoyed at this recognition of his authority, and exultantly greeted him: "Peace on thee, my master and my disciple; my master in learning, and my disciple in acknowledging the authority of my office."[8]

On another occasion, Rabbi Joshua and the patriarch clashed over a question of ritual procedure. Rabban Gamaliel ruled religious worship in the evening obligatory. It was an innovation in tradition, but he apparently judged it necessary because the Temple had been destroyed and formalized prayer was to replace the cult of sacrifice as an organized and authoritative expression of Jewish piety. Rabbi Joshua wanted more spontaneity in religious life, and to a student who had consulted him, he expressed himself that the evening service ought to remain voluntary. The patriarch heard of this, and he decided to make another demonstration of his authority. When the Sanhedrin gathered for a formal session, Gamaliel had the question submitted for formal consideration, and then he repeated his ruling that the evening service is obligatory, asking whether there were any dissenting opinions. Rabbi Joshua, who was chief justice, announced that there were none. Whereupon Rabban Gamaliel ordered, "Joshua, stand up and a witness will testify that you dissented." Rabbi Joshua confessed his guilt, but the patriarch, as a mark of displeasure, left him standing throughout the day's proceedings.

There was shock among the rabbis at the overbearing and dictatorial manner of the patriarch. When the memorable session broke up and Rabban Gamaliel departed. the members of the Sanhedrin reassembled. After gravely considering the difficult situation that had developed, a motion was made and carried impeaching the patriarch from his office as head of the academy. He remained patriarch but he was shorn of the academic prerogatives which had gone with the office. A younger member of the court, Rabbi Eleazar ben Azaryah was elected head of the Academy. Rabban Gamaliel accepted the verdict calmly and took his place as a lay member of the court, bearing no grudge and carrying no vindictiveness. Duly humbled, moreover, he apologized to Rabbi Joshua for having treated him with discourtesy. A reconciliation followed and Rabban Gamaliel was finally reinstated, but since Rabbi Eleazar had held the high office, he was to share some of the prerogatives of the office with Rabban Gamaliel. Thus he was to deliver the public lecture on the Sabbath every third week.[9]

REBELLION RENEWED

The program of reconstruction as inaugurated at Jabneh was interrupted by a new uprising against Rome. Scattered remnants of the old army of zealots who had challenged Rome in 70 C.E. fled Palestine to settle in various other centers of Jewish population within the empire, including Egypt, North Africa and Cyprus. There they had sown the spirit of discontent and rebellion. And in 116 when Trajan launched a campaign of new conquest in the East, the Jews renewed the old struggle. Palestine played a minor role in the uprising. Jabneh exerted an influence of moderation. the rabbis seeking to dissuade their people from resuming the struggle on the military level. But the rebellion was pursued with unprecedented bitterness and determination in the Jewish diaspora. Fierce battles raged in such cities as Alexandria, Cyrene and in Cyprus, with casualties running in the hundreds of thousands. But it was not a clear-cut struggle between the Jews and the Romans. For many of the natives in each of these conquered provinces bore willingly the yoke of Roman imperialism. And when the crisis was precipitated the native collaborators of the Romans struck out against their Jewish populations who alone battled for freedom. The civil wars which thus ensued doomed the rebellion to failure. There were new executions, new persecutions, and new despair. Most of the Jewish communities in the diaspora were destroyed. But an embassy of rabbis under the leadership of Rabban Gamaliel hurried to Rome and succeeded in warding off a series of retaliatory measures which had also been projected against Palestine.

A new storm broke after Hadrian came to power (117–138). Content to mark time within the old frontiers and temporarily not to encroach upon the domains of the sturdy Parthians, who maintained a free and flourishing kingdom in what is present-day Iraq, Hadrian became a reformer, devoting his energy to the inner needs of his vast empire in preparation for the next Roman bid for world conquest. One of the most vital of these needs, as he saw it, was the strengthening of imperial unity which he tried to achieve through the cultivation of a common culture. What culture other than Hellenism enjoyed the prestige qualifying it to become the imperial culture? The experiment of Antiochus IV, King of the Syrians, was to be tried again, though on a minor scale. The heterogeneous territories of Rome were to be molded into a national state, linked together by the culture of old Hellas. His program included the restoration of Jerusalem, but as a pagan city, its crowning edifice to be a Greek Temple dedicated to Jupiter.

There was consternation among the Jews when this program was announced. For it held out its threat not merely to Jewish social and political institutions, but to the Jewish way of life, to the Jewish religion. With the Torah in jeopardy, the rabbis now joined the camp of open rebellion. The youthful but brilliant Talmudist, Rabbi Akiba ben Joseph, a leader of the Jabneh Sanhedrin, gave his blessings to a new anti-Roman rebellion which was proclaimed by Bar Kokba. The new insurrection broke out in 132, and it registered some initial success. For 2½ years the Jews held the recaptured city of Jerusalem. They even made attempts to restore the sacrificial cult at an improvised altar. Coins were struck proudly marked in honor of the First, Second and Third Year after the Liberation of Jerusalem. The Romans, however, soon reasserted their power. Jerusalem was retaken. The Jews entrenched themselves in the fort of Bethar, southwest of Jerusalem, but were forced to surrender in 135. Half a million Jews are said to have perished in the struggle. Judea was practically turned into a desert; its cities and villages were in ruins.

The Romans had paid dearly for their victories. So huge were the Roman losses that the emperor omitted the usual reassuring formula from his report to the Senate: "I and my army are well." But they could at last proceed with their plans. Jerusalem was plowed up to dramatize the new foundations of the city, to be called Aelia Capitolina. Temples were built to Bacchus, Serapis, Venus and Jupiter. No Jew could set foot into the new city. The practices of Judaism were forbidden on the pain of death. There was to be no observance of the Sabbath, no performance of the rite of circumcision, no study of Torah, and, to break the continuity of an authoritative religious leadership, they outlawed the ordination of new rabbis. Stricken at the source of its vitality, this dissident people was at last to give way and the totalitarian empire was to pursue unchallenged its ambitions of building a new world state.

In the contest of arms, Rome had once more emerged victorious. But brute power, no matter how overwhelming, has generally proven impotent in the face of a people that was actuated by a profound will to live, and was prepared to pay the cost of survival in suffering. There were Jews. in all layers of society who no longer had the strength to suffer and whose morale was waning. Elisha ben Abuyah, a famous figure in the rabbinical academies, some of whose moral maxims have been preserved in the ethical treatise Abot, turned renegade and offered himself as a willing collaborator to the Roman officialdom. He helped the Roman police in ferreting out leaders of Jewish resistance and in closing schools where the old way of life was being taught. Rabbi Jose ben Kisma who had once claimed, "If all the precious metals in the world were offered me, I would not live but in an atmosphere of Torah," saw in the repeated successes of Rome the evidence of divine favor. Rome was apparently invincible and it was folly to resist the sweep of the future.[10]

But there were others, of sterner stuff, who did not lose themselves in the crisis. Rabbis Akiba, Tarfon, and Jose the Galilean held a secret conclave and issued a joint statement to their people, urging them generally to comply with Roman edicts but to resist unto death any orders involving the commission of idolatry, murder, or unchastity.[11] And a company of distinguished teachers openly defied the Roman police by continuing to meet with their students for the study of Torah. Their attitude was best summarized in Akiba's famous parable of the fishes and the fox. Warned that his open defiance of Roman law would lead to imprisonment, he replied with the story of the fox who invited the fishes to seek safety from the fishermen on dry land. But the fishes replied, "If the water which is our normal habitat hold out no safety, what will happen to us on the dry land which is not our habitat?" "So, too," expounded Akiba, "if our existence is precarious when we persist in the study of Torah, how shall we survive if we abandon it?"[12]

The forebodings came true soon enough. The Romans unleashed a reign of terror against the obdurate Jews; many were imprisoned, banished or sold into slavery. There were numerous executions. Intimidated by the terror, many fled Palestine to neighboring countries, particularly Babylonia, where a more tolerant government offered these political refugees a ready welcome. The records of some of those who perished in the terror have been preserved and they recount a memorable story of steadfast faith and heroic struggle. The Midrash Eleh Ezkerah is the vivid description of a mass execution of ten renowned rabbis. It has been rendered into verse and included in the liturgy of the Day of Atonement.

Among those arrested by the Romans was Rabbi Akiba. From his prison cell, he continued to defy his captors, dispatching secret messages to his followers. A hurried trial was held and the Romans condemned him to death. According to tradition, they executed him by tearing the flesh from his living body. Rabbi Akiba remained steadfast to the very last, expiring with a resolute confession of his outlawed faith: "Hear, O Israel, the Lord our God, the Lord is One."[13]

Akiba's work was immediately taken up by Rabbi Judah ben Baba. Gathering Akiba's five most gifted disciples, Meir, Judah ben Ilai, Simeon ben Johai, Jose ben Halafta and Eleazar ben Shammua, he officially conferred upon them rabbinical ordination and charged them with the task of continuing the tradition of a courageous and devoted leadership amidst confusion and terror. The meeting was raided by the Romans before its conclusion. The five younger men were able to escape, but Rabbi Judah ben Baba was stabbed to death.[14]

THE RABBIS AT THE HELM

Hadrian's experiment in totalitarianism came to an end with his death in 140, when he was succeeded by Antoninus Pius. The Sanhedrin was at once reorganized, but it now abandoned Jabneh for Usha in Galilee. As one Midrash relates it, "At the termination of the persecutions, our teachers met in Usha. They were Rabbis Judah ben Ilai, Nehemiah, Meir, Jose, Simeon ben Johai, Eliezer (the son of Rabbi Jose the Galilean), and Eleazar ben Jacob. They sent to the elders in Galilee saying, 'Those who have already learnt, come and teach; those who have not yet learnt, come and be taught.' They met and arranged everything that was necessary."[15]

One of the things which they finally arranged was the selection of another scion of the Hillel family as their leader, Simeon, the son of Gamaliel II. Simeon, the new nasi, had been in Bethar, the last Jewish stronghold, during the Bar Kokba rebellion, and had witnessed its fall to the Romans. A narrow escape saved him from the massacre that followed in the city.

Having experienced the horrors of war, Simeon extolled the virtues of peace. "The fabric of civilization rests on three moral foundations—truth, justice and peace." "Great is peace, for Aaron the High Priest acquired fame only because he promoted peace." He is also the author of the famous maxim: "It is unnecessary to erect monuments to the righteous; their deeds are their monuments." Rabban Simeon advocated equal justice to Jews and pagans. In one instance he declared it obligatory to ransom pagan slaves who had been kidnapped. And he was lavish in his admiration for those semi-Jews, the Samaritans, for loyally observing those elements in the Torah which they recognized. He was so highly esteemed by the Sanhedrin that in all but three instances his opinion was accepted as authoritative law.

The achievement of the rabbis at Usha extended over a varied field and continued the precedents established at Jabneh. They defended the supremacy of Palestine as the center of Judaism against the claims of the rising Jewish community in Babylonia. An attempt had been made to found a Sanhedrin in Babylonia during the Hadrianic persecutions of Judaism in Palestine. It threatened to start a schism in Israel. When the Usha Sanhedrin was reorganized, a delegation of two rabbis was sent to Babylonia and they succeeded, after a struggle with the local authorities, in inducing the Babylonians to continue to heed Palestinian leadership. They reorganized the procedure of the Sanhedrin to invest its sessions with new pomp and dignity. In addition to the nasi and chief justice, a new office was created, the *hakam*, literally a "wise" or learned man. The new functionary seems to have been in charge of the academic functions of the Sanhedrin.

They continued the process of literary reorganization in traditional law as well as the application of that law to new situations; and they enacted a number of reforms dealing with various aspects of religious, domestic, and social life. They ordained that parents were to maintain their children throughout their minority, and that where parents deeded their property to their children, they must be supported from the estate; a person was to contribute a fifth of his income to charity; a father must be patient in teaching his children till the age of twelve, but thereafter he may take severe measures with them.16 Perhaps the most important reform was the declaration of immunity for members of the Sanhedrin who could not be excommunicated for their views, regardless of circumstances.

Judah I, who succeeded his father Simeon to the patriarchate, was born about 135. He received his education at Usha under his father and from intimate contact with the various members of the Sanhedrin. It is uncertain when he assumed the office of his father or when the seat of the Sanhedrin was transferred from Usha to Bet Shearim, also in Galilee. He was not of very good health and for the last seventeen years of his life he lived in Sepphoris, a section of the country renowned for its high altitude and pure air. He died in 217.

Judah was a man of very great wealth and was held in high esteem by Jews as well as by Romans. His universal recognition as a master of tradition and leader in Israel is well attested by his popular designation "Rabbi," without his name, the master par excellence, or Rabbenu ha-Kadosh, "our holy master." The Sanhedrin over which Judah presided had no chief justice or hakam; he himself fulfilled the varied functions for which the other two offices had been created.

A number of his aphorisms have been preserved in the Talmud. "I have learned much from my masters, more from my colleagues, but most of all from my pupils." "Do not consider the vessel but its contents; many a new vessel is full of old wine and many an old vessel is without even new wine." To the question, "Which is the right path that a man is to choose in life?" he offered the following answer: "That which will be a source of pride to him, before his own conscience, and which will also bring him honor from mankind." He held that children must revere both parents equally.

Under Judah's leadership the Sanhedrin enacted a number of important social reforms. Those who purchased property twelve months after its seizure by the Roman government were required to compensate the original owners by a fourth of its purchase price, but the transfer of possession was declared valid. Certain frontier districts of Palestine were exempted from tithing their crops or leaving their land fallow on a sabbatical year. The theory on which these exemptions were made was that those had not been part of the country regions originally invaded by the Israelites under Joshua. But their purpose was obviously to alleviate conditions among the Jewish masses upon whom the tithes and the seventh year loss of the harvest had proven a very heavy burden.

Rabbi Judah sought to abolish the fast of the ninth day of Ab, when Jews mourned for the fall of the Temple. There was no point in maintaining the fast, he felt, since the Jews were free of persecution and were living everywhere as a free community within the Roman Empire. Indeed, the continued commemoration of that fast day fostered ill-will between Jews and Romans. But Rabbi Judah's colleagues opposed the move and the fast remained.

The most important achievement of Judah was the completion of the great literary enterprise that had been started in Jabneh—the compilation of the Mishnah. Judah synthesized in his work all that had been accomplished before him. He relied particularly on the compilations of Rabbi Akiba and Rabbi Meir. But he ultimately made them all his own and what he produced was a succinct and comprehensive record of Jewish legal tradition from the dawn of Pharisaism until his own times. The opinions that he recognized as authoritative law are generally presented anonymously, but the views of the dissenting masters are given as well. With some minor variations, the product that left the hand of Judah is the classic text of the Mishnah which has been preserved to our own day.

With the passing of Judah I, the old lustre departed from the office of nasi. A number of more or less inconspicuous personalities succeeded him, but they made little mark for themselves as teachers and leaders in Israel. The most important of them was Judah's grandson, Judah II. Judah was a close friend of the Roman Emperor Alexander Severus. To facilitate more cordial relations between Jews and Romans, he sought to abolish some of the restrictions against free relations with pagans. He succeeded in lifting an old ban against the use of oil bought from pagans. This was also an important economic amelioration for the Jewish community which used oil as a staple in the common diet. In his own home, the patriarch allowed himself certain deviations from Jewish custom, yielding to the influence of Roman manners to which he was freely exposed. He was openly criticized for this, but the rabbis rationalized that as representative of Jewry he was obliged to mingle with Roman officials which made such accommodations inevitable.[17]

The rabbis who functioned after the compilation of the Mishnah were called *amoraim*, expositors, to distinguish them from their predecessors who were called *tannaim*, teachers. The Mishnah had greatly simplified their labors, for they now had an authoritative record of tradition on which to base their interpretations and decisions.

The pioneer of amoraic activity was Rabbi Johanan (199–279), who headed the academy in Tiberias and who has frequently been called the compiler of the Palestine Talmud. Johanan had studied under Judah I and he extolled the value of his Mishnah. "I base all things on the Mishnah," he declared. Johanan established the principle that no amora could contradict a tanna unless he had tannaitic support for his position. "Whatever is written in the Mishnah has been communicated to Moses on Mount Sinai."

Six commandments he extolled with particular emphasis: hospitality to strangers, visiting the sick, careful prayer, rising early to go to the academy, raising children to the knowledge of the Torah, and judging everyone according to his good deeds. Johanan was a great humanitarian. He treated his slave as an equal and served him regularly the same food eaten by the rest of the household. "The slave," he explained, "is the same child of God that I am." He suspended all laws proscribing labor on the sabbath to save a sick person who could then live to observe many sabbaths. He ruled that the injunction to return a straying ox or sheep (Deut. 22:1) applied even if the owner was a Jew who had renounced his Judaism, and he called upon people to give full recognition to whatever truths pagan wise men might discover.

He complained bitterly about the oppressive taxations levied by the Romans. "Such is the way of an evil kingdom when it proposes to seize people's property," he once explained. "It appoints one to be an overseer and another a tax collector. By these devices it takes away the possessions of people." He was hopeful that the Parthians would finally prevail and make good their challenge to Roman supremacy, thereby liberating Jewish Palestine. But, in spite of prevailing persecutions, he sought to discourage the emigration of Jews from Palestine. The national cause demanded that Jews hold on to every position in their native land.

Johanan did not edit the Palestinian Gemara, as has sometimes been asserted. Johanan died in 279, and the Palestinian Gemara quotes scholars who lived in the fourth and early fifth centuries. But Johanan's work was the most important contribution to the making of this Palestinian supplement to the Mishnah.[18]

After the passing of Johanan, there set in a continuous decline in Palestinian scholarship. The conversion of the Roman Empire to Christianity under Constantine (311–337)—another experiment in unifying a heterogeneous empire with a common faith and a common way of life—brought new disabilities upon the Jewish community. In 351, the Jews attempted another uprising against Rome and brutal retaliations followed. The most important centers of Jewish life and learning, including Tiberias and Sepphoris, were destroyed. The Roman Empire itself was weakening, moreover, under the weight of constant warfare with the Parthians and neo-Persians in the East. In the early fifth century, the west was invaded by the Goths and Vandals, and chaotic conditions spread throughout the empire. In Palestine, as in the other provinces, the population was constantly diminishing, through natural decline, as well as through emigration and social and cultural life gradually came to a standstill.

The decline of Palestinian Jewry is perhaps best illustrated by the patriarchs who succeeded Judah II. They were essentially figureheads as far as their functions in Jewish life were concerned. Their knowledge of tradition was mediocre. They were the political representatives of Jewry before the Roman officialdom and assisted the Government in the collection of taxes. The leadership in cultural life passed into the hands of the amoraim, who carried on their work more or less independently, without centralized direction from the patriarchal office.

As nominal heads of the Sanhedrin, the most important function of these patriarchs was the annual promulgation of the calendar. Even this ceased in 359 when the patriarch, Hillel II, formulated a mathematically calculated fixed calendar, doing away with the periodic calendar determination on the basis of the observed position of the new moon. The inertia of a long and colorful past kept the office going for another sixty years, but its usefulness had long since ended. And in 425 when the patriarch Gamaliel VI died childless, the patriarchate was officially abolished.

The abolition of the patriarchate marked the termination of Palestine's role as a center of Judaism. We do not know whose hands put the finishing touches upon the literature of Mishnah supplementation, the Gemara, which had grown up in the various schools. Together with the Mishnah, it proved to be a rich legacy that a fruitful and creative epoch had left to its posterity.

A NEW DAWN IN BABYLONIA

Babylonia, the modern Iraq, ranks second only to Palestine as a center of classical Judaism. Situated along the lower reaches of the Tigris and Euphrates rivers, the country is rich in alluvial soil, and was one of the most fertile regions of the ancient world.

There had been a Jewish community in Babylonia ever since the Babylonian King Nebuchadnezzar had destroyed the independence of Judah in 586 B.C.E. and deported to Babylonia a large part of the Judean population. including the leaders of political and religious life. The deportees were given their freedom and were allowed to settle on the land or to engage in any other pursuits of their liking. Babylonia's rich soil rewarded their labors with a lavish bounty; and they grew and prospered in the new land.

The political convulsions of the ancient world repeatedly bore their full weight upon Babylonian life. Cyrus, the Persian, conquered the country in 539 B.C.E., and his dynasty maintained its domination for more than two centuries. Persian power was broken by Alexander the Great in 331 B.C.E. In the division of empire which followed Alexander's death, Babylonia was joined to Syria as part of the kingdom falling to the general Seleucus.

In one essential respect did the fate of Babylonia differ from the rest of the Near East: she never succumbed to Roman conquest. In 160 B.C.E., a Parthian King, Mithridates I, established himself in the country and the new dynasty, which reigned until 226 C. E., built a powerful military machine that was repeatedly able to hurl back the Roman legions seeking to invade it.

The Jews suffered all the repercussions of the various wars that ravaged the country, but their freedom remained intact. And with the benevolent cooperation of the government, they evolved an ingenious pattern of community life. The Jews were recognized as a national minority, governed by a hereditary prince, called the *Resh Galuta*, the head of the captivity. This prince who was a descendent of the royal house of David, was the fourth highest ranking noble of the state, representing the Jewish population. As far as the Jewish community was concerned, he was vested with the right to supervise trade and commerce, to appoint judges and to direct the various other tasks of regional government. including the collection of taxes.

The autonomous organization of the Jewish community was made possible by the compact character of the Jewish settlements. Such cities as Nehardea, Nares, Sura, Mehoza, and Pumbedita, had predominantly Jewish populations. The contact between Jews and native Babylonians was free and unrestrained; and the impact of mutual influence was evident on both cultures. The Jews adopted the Aramaic vernacular spoken in Babylonia. Many Babylonians, on the other hand, embraced the Jewish faith and were welcomed into the synagogue. Among the greatest triumphs to Jewish proselytising in Babylonia was the conversion of the royal family of Audiabne, a vassal state in northern Mesopotamia.

Pursuing their own culture, the Jews of Babylonia had developed important centers for the study of the Torah. The pioneer of the sopheric movement in Palestine, Ezra, had received his training in Babylonia. The venerable teacher, Hillel, received his preliminary education in the Babylonian schools, before migrating to Palestine. The chief justice of the Palestinian Sanhedrin under Simeon ben Gamaliel III was a Babylonian scholar, Nathan. a member of the family of the Resh Galuta.

But the Jews of Babylonia realized that the central authority in Jewish life must be directed from Palestine. Most of them did not want to leave their new homes to take advantage of the Persian edict allowing them to return to Palestine. But they organized an expedition of pioneers who were willing to return; and they helped the new Palestinian settlement with material and moral support until it could get on its feet to resume once again a normal national life. And when Palestine was prepared to offer leadership, they gladly followed. They sent their annual contributions for the maintenance of the Temple and went on pilgrimages to Jerusalem. They respected the authority of the Sanhedrin and its hierarchy of teachers and leaders who directed Jewish religious and cultural life.

The Roman invasion sent a new wave of immigration from Palestine to Babylonia. This was augmented particularly after the Jewish uprisings of 70 C.E. and 135 C.E. when the Romans devastated Judea and destroyed the Temple. As a Babylonian teacher, after witnessing the tragedy of Palestine, under Rome, remarked: "The Lord, knowing the Jews would not be able to bear the hard decrees of Rome, exiled them to Babylonia."

Among these new Palestinian emigres who came to Babylonia was Hananiah, a nephew of Rabbi Joshua ben Hananiah, and also a distinguished scholar. Apprehensive that the Romans would completely uproot Palestinian Jewry, Hananiah attempted to organize a Sanhedrin in Babylonia. He organized a school at Nehar Pekod for advanced study of the

Torah, and, as its head, proclaimed his competence to promulgate the calendar of Jewish festivals, without reference to the Palestinian authorities. He won local Jewish support, and Babylonian Jewry thus seceded from its dependence on Palestine.

Hananiah's attempt was, however, premature. Babylonia's time had not as yet come. For in 140, peace was restored in Palestine and the Sanhedrin reconstituted at Usha. To recover the loyalty of Babylonia, the Usha Sanhedrin sent a delegation of rabbis to Hananiah, who, however, remained obdurate. The delegates finally appealed, over Hananiah's head, to the Jewish laity.

The Talmudic story relates that one of these delegates when called upon to read the Torah at a synagogue service on a festival date fixed by Hananiah's calendar, substituted "These are the holidays of Hananiah" for "These are the holidays of God." Members of the congregation, of course, corrected him, but he replied, "It is we in Palestine who may read, 'These are the holidays of God'; here in Babylonia one must substitute the name of Hananiah since he fixes the holidays as he chooses and not as God commanded." The second delegate then arose to read and he recited the verse, "Out of Zion shall go forth the Torah," as "Out of Babylonia shall go forth the Torah." When corrected, he replied similarly: "In Palestine we may read as written, but judging by your conduct the amended reading appears justified." Public pressure finally forced Hananiah to yield; and Jewish life was once again under a unified centralized leadership, the Palestinian Sanhedrin.[19]

The Jews in Babylonia continued to heed the Palestinian Jewish leadership until the fifth century when the Sanhedrin was abolished. But as Jewish life in Palestine declined and Babylonian Jewry grew with ever more accretions from Palestine, the active enterprises of Jewish culture were increasingly transferred to Babylonia. By the third century, a series of schools of higher Torah studies sprang up in Babylonia which conducted for centuries an active intellectual life, the choicest product of which is the Babylonian Gemara.

The pioneer in this intellectual development was Rabbi Abba Areka, or as he was popularly designated *Rab*, master par excellence. He had studied in Palestine under Rabbi Judah I, where he distinguished himself in his studies, winning the plaudits of a scholar like Rabbi Johanan who acclaimed him as his own superior. It was at Judah's school that he probably acquired the epithet Areka, tall, to differentiate him from another scholar by the same name, the father of the famous Babylonian amora, Samuel.

In spite of his high standing as a scholar, Rab could not achieve full ordination from Judah or his successor, Gamaliel III. The patriarchs were apparently unwilling to invest Abba with full authority because they did not want to see a self-sufficient religious life established among the Babylonian Jews, with a rival academy under Abba's leadership. The hegemony of Palestine remained unbroken so long as it remained the only Jewish community with a fully ordained religious leadership. Nevertheless he was recognized as an equal in rank to the great teachers who were active before the compilation of the Mishnah and privileged freely to dispute their opinions, a distinction accorded to him alone among all the amoraim, as the rabbis of post-Mishnaic times were called.

Rab returned to Babylonia in 219 with a widely recognized reputation as a great scholar. The Resh Galuta at once appointed him commissioner of markets in Nehardea and he was invited to lecture in the local academy which was then headed by a certain Shila. But neither office proved an adequate outlet for his independent spirit and creative intellect. When Shila died, he was offered the rectorship of the Nehardea academy, but he turned it down, with the recommendation that it be offered to a local scholar, Samuel, who had a greater claim to it.

Rab's final decision was to become a pioneer, to found a new academy with new traditions and in a region where there was little knowledge of the Torah and a widespread neglect of Jewish religious life—the city of Sura. As the Talmud puts it, he entered "an open and neglected field and fenced it in." With his own funds he erected a school building, and he offered scholarships to needy students, providing tuition and maintenance, and before long 1200 students had enrolled to study under him.

Rab had brought with him from Palestine the text of the Mishnah, edited by Judah I, and he based all his lectures on it, supplementing it, however, with explanations, illustrations, and various new applications. But he was equally interested in the exposition of moral lessons. The Talmud has preserved a number of his moral maxims and they are among the choicest ethical expressions in all literature: "The rituals of the Torah were given only to discipline men's morals." "Whatever may not properly be done in public is forbidden even in the most secret chamber." "It is well that people occupy themselves with the study of Torah and the performance of charitable deeds even when inspired by ulterior motives; for the habit of right doing will finally ennoble their intentions as well." "Man will be held to account for having deprived himself of the enjoyment of good things which the world has offered him." "When necessary, flay dead carcasses in the street and do not say, I am a priest, I am a great man." "It is better to throw one's self into the fiery furnace than to humiliate one's fellow-man."

Rab did not confine his interests to the classroom alone. He was one of the most active and influential communal leaders in his day. He enriched the Jewish liturgy with a number of beautiful prayers which still offer the most inspiring motifs of the synagogue service. His compositions include the prayers on the occasion of the new month, a good deal of the New Year and Day of Atonement liturgy, and the Adoration, Alenu Leshabeah, with which every Jewish religious service, private or congregational, is concluded.

He interested himself in elementary education, ordaining that pupils shall not begin their studies before the age of six; that teachers must create discipline through the winning of the interests and affection of children rather than through corporal punishment; and that no children shall be deemed unworthy of instruction because of mental backwardness. He also contributed immeasurably to the reformation of Jewish family life. Thus, he ruled against child marriages and advocated a period of courtship to enable the boy and girl to discover their own preferences and to choose their own mates without parental dictation.

Rab started his school at the age of 64; and he continued as its active head for 28 years. When he died, all the Jews in Babylonia mourned him for a full year, observing all the rites of mourning which are followed at the loss of a member of the family. He had made himself immortal by raising the religious and cultural life of Babylonian Jewry and by establishing a great institution of high learning which was destined to endure, with minor interruptions, for seven centuries.[20]

The academy at Nehardea which had invited Rab to its leadership after the passing of Shila was presided over by the local scholar, Samuel, or as he was often called, Mar (Master) Samuel. Like Rab, he had studied in Palestine, but he did not receive even partial ordination. In addition to his studies of Jewish tradition, he had an excellent scientific training. He was a practicing physician and a well-known astronomer. It is interesting that he traced many diseases to the unhygienic conditions under which people lived. He was especially famous for his skill in treating the eye. He denounced the then, and, in many circles, still prevalent superstition ascribing diseases to the evil eye. He attributed a great deal of therapeutic value to air and climate.

As head of the Nehardea academy, Samuel distinguished himself particularly in the field of civil law. He was the author of the famous principle of Jewish law that the legal system of any country where Jews are residing is binding upon them, even when it is in conflict with their own system of law. And though the Babylonian Jews enjoyed an autonomous court system, he demanded that Jewish judges reckon with the prevailing Babylonian law in reaching their decisions.[21]

Between Rab and Samuel there was constant intellectual commerce, and the two men frequently disagreed. Samuel's leadership was followed to the full in his specialty, civil law, while Rab remained supreme on questions of ritual law. Samuel survived his colleague by ten years, and during that time Sura was without a successor to Rab. The Nehardea academy was looked upon as the supreme center of Jewish scholarship and religious authority in Babylonian Judaism. The city of Nehardea was sacked in 261 as one of the incidents in the constant warfare between the Babylonians and the Romans. But the famous academy was transferred to a neighboring city, Pumbedita, by a pupil of both Rab and Samuel, Judah ben Ezekiel. This famous academy now remained in Pumbedita throughout the Talmudic period, except for a brief interlude between 339 and 352 when the new rector, Raba, had it transferred to his native city, Mahoza.

The Babylonian academies developed an ingenious educational institution which enabled them to reach large numbers of non-professional students. During the two months, Adar and Elul, when the average farmer is free from his work in the fields, special educational sessions, called *Kalla* were held in the schools of Sura and Pumbedita. The subject to be taken up at each of these sessions was announced in advance, and laymen were encouraged to spend their hours of leisure in preparation. The lectures of the rector of the academy were supplemented with the discourses of special lectures. The basic text discussed in all these gatherings was the Mishnah; and one tractate was generally covered each month. 12,000 students are reported to have been enrolled in one such Kalla session.[22] This institution brought the academies into close contact with life, and closed the gap between the professional scholar and the layman. And more than anything else, it helped raise the cultural level of Babylonian Jewry.

The two academies continued their parallel development, under various heads, until the end of the fifth century. There were other schools similarly dedicated to the advanced study of the Torah, but Sura and Pumbedita were the supreme centers of intellectual life, where the Mishnaic utterances, so succinctly formulated by Judah I, were enriched with a supplement of new legal discussions and where the doctrines of Jewish theology and ethics were expounded to offer inspiration and guidance to a new community in Israel.

The turning point in this cultural enterprise was the new domestic policy adopted by Kings Yezdegerd II (438–57) and his son Peroz (459–86). Celebrating a respite from the constant warfare with the Romans, these kings entered upon a policy of the intensive nationalization of their realm. Zoroastrianism as the official religion of the country was proclaimed the medium of national integration as well, and all dissident religions were proscribed as treason to the state. The blow of the new nationalism fell with particular severity upon the Jews. The observance of the Sabbath was prohibited. Synagogues were destroyed and schools closed. Jewish children were caught and delivered to the priests of Zoroaster to raise them as devotees of the national religion. Under these severe persecutions, the once proud Jewry of Babylon began to crumble. There began a mass flight of Jews to friendlier shores, to Arabia, India and the Caucasus.

The accumulated cultural achievements of the Babylonian academies were, however, preserved for posterity through the timely labors of principally two men, Rab Ashi (d. 427) and Rabina (d. 500). Ashi headed the Sura academy for more than 50 years of peace, and he had begun the organization of the vast literature which had grown up in the different academies, around that unique classic of Jewish tradition, the Mishnah. What Ashi started as a leisurely work of detached scholarship became a pressing necessity after his death when the Jewish community was overwhelmed with a great disaster. And his successors at the Sura academy continued his work, finally completing it and, it seems, also reducing it to writing. The final job of editorial revision was rendered by Rabina, the last of the rectors of the Sura academy in Talmudic times. Babylonian Jewry eventually recovered from these persecutions and the schools were reopened for another five centuries of cultural leadership. But the *Gemara* which had been completed by the close of the fifth century was the apex of its cultural life and its chief contribution to the Judaism of the future.

THE THEOLOGICAL ELEMENTS IN THE TALMUD

THE EXISTENCE OF GOD

The Talmudists, like their Biblical predecessors, assert the existence of God, and from the perspective of that assertion interpret all life in the universe. They do not engage in any formal efforts to prove God's existence. Yet there are passages in their writings which show that they could follow the logic of the philosophers and reason from nature to a first cause. Such reasoning is put by the rabbis into the mind of Abraham as they trace the odyssey of his own faith, from idolatry to monotheism. According to one account Abraham inferred the existence of God by contemplating the universe as one may infer the existence of some master when viewing a palace brilliantly illuminated within. "Can it be that the universe and all that exists within it is without a directing mind?" Abraham is quoted as speculating. The universe in itself did not, however, answer Abraham's quest. God met him halfway, and rewarded his groping by revealing Himself to him with the reassuring word of His presence. "The Lord looked upon him and said: 'I am the master of the universe.'"[1]

The assertion that the universe is the creation of God does not make clear the many varied and intriguing problems that the contemplation of existence presents to alert minds. What, for instance, was before creation? And how did creation itself really transpire to fashion a universe out of nothing? But the Talmudists discouraged the preoccupation with such problems. They held the ultimate mysteries of the world beyond human comprehension, and they felt that the concentration upon them, a futile enterprise, in the long run, would only have the immediate effect of distracting men's minds from the more pressing tasks of religious and moral duty.

Their apprehensions were reinforced by the tragic experiences of the famous four teachers who had studied the ultimate mysteries: Akiba, Ben Azzai, Ben Zoma and Elisha ben Abuyah. A cryptic passage tells of their fate: "Ben Azzai gazed and died; Ben Zoma gazed and became demented; Acher (Elisha) cut the plants (turned apostate); R. Akiba departed in peace." The Talmudists therefore warned their people with the well-known citation from Ben Sira: "Seek not out the things that are too hard for thee, and into the things that are too hidden inquire thou not. In what is permitted to thee instruct thyself; thou hast no business with secret things."

The distinction between "what is permitted" and the "secret things" which are not permitted, is set forth in the Talmudic observation as to why Scripture commences with the second letter of the Hebrew alphabet, the Beth, rather than with the first. The explanation is seen in the symbolism which that letter's shape, a square open on the left side, suggests: "As the letter *beth* is closed on all sides and only open in front, you are to regard as closed to inquiry what was before creation or what is behind; what is open begins from the actual time of creation."[2]

The rabbis, following the style of the Bible, frequently spoke of God as though He were a person. They ascribe to Him bodily attributes. It is clear however, on the basis of their own declarations, that these "corporeal" references to God were often intended only to make vivid the sense of His existence and activity. All such references are to be taken as figurative expressions. Even the story of God's revelation at Sinai is taken in the Talmud by one teacher in a figurative way. "Moses never ascended to heaven," declared Rabbi Jose and "God never descended on earth." The Biblical narrative is to be taken as a poetic elaboration of the doctrine that God was the inspiration for the truths which Israel pledged itself to uphold at Sinai. It must not be taken literally.

The rabbis insisted repeatedly that God is not a concrete being, with tangible form, occupying a specific magnitude in space. Such a being would be part of the universe, not its master. Indeed, one of the epithets by which God is referred to in the Talmud is "The Place", for God is the "place" or the ground of creation; the universe exists in Him not He in the universe. In the words of the Midrash "The Holy One, blessed be He, is the place of His universe, but His universe is not His place."[3]

But by a paradox of the divine mystery God, though transcending the universe, is yet ever present, and men can enter into close and intimate communication with Him, wherever they are. This is the significance of the revelation of God perceived by Moses at the burning bush—it is to teach us that there is no place which is devoid of God's presence, not even so humble an object as a thorn bush. Another rabbi declared: "At times the universe and its fulness are insufficient to contain the glory of God's presence; at other times He speaks with man in intimate discourse."[4]

The assertion that God is invisible made him unreal for people accustomed to identify reality with concreteness. But the rabbis disputed this. Thus it is related in a Talmudic anecdote that the Emperor Hadrian had said to Rabbi Joshua ben Hananiah: "I desire to behold your God." Rabbi Joshua explained to him that it was impossible. When the emperor persisted, the rabbi asked him to stand in a fixed gaze at the sun. The emperor found the sun too strong. Thereupon the rabbi exclaimed: "You admit that you are unable to look at the sun, which is only one of the ministering servants of the Holy One, blessed be He; how much more beyond your power of vision is God Himself." Rabban Gamaliel explained the reality of God by analogy to the soul whose specific abode we do not know and of which we have no direct concrete experience. That, however, does not make it unreal.[5]

God's ultimate essence must elude human comprehension. We may, however, see manifestations of divine activity throughout creation.

The Talmudists saw a manifestation of God in the dynamism of the world. The universe is not a mass of inert matter. It is an enterprise of tremendous dynamic activity. "The universe is filled with the might and power of our God.... He formed you and infused into you the breath of life. He stretched forth the heavens and laid the foundations of the earth. His voice blows out flames of fire, rends mountains asunder, and shatters rocks. His bow is fire and His arrows flames. His spear is a torch, His shield the clouds, and His sword the lightning. He fashioned mountains and hills and covered them with grass. He makes the rains and dew to descend, and causes the vegetation to sprout. He also forms the embryo in the mother's womb and enables it to issue forth as a living being."[6]

In this vast panorama of existence, moreover, there is the evidence of a purposeful Intelligence at work. No organism is superfluous. A close scrutiny of the world shows everywhere an all-permeating intelligence and purpose. We see the evidence of that design in the vastness of the planetary system, in the individuality of each rain drop, in the majesty of trees that renew their garb of green in spring, in the mysteries of love which bind men and women in the unity of marriage. "Even such things as you deem superfluous in the world, such as flies and gnats are necessary parts of the cosmic order and were created by the Holy One, blessed be He, for His purpose—yes even serpents and frogs." Indeed, every creature in its own way, by its mere existence, and by the precision with which it functions in the world, offers eloquent testimony to the divine source from which it is derived.[7]

It is in man that the design of creation shows itself most forcefully. The Talmudists admired the marvellous construction of the human body in which every organ seemed so perfectly designed for the well-being of the individual and the furtherance of life. "Come and see how many miracles the Holy One, blessed be He, performed with man, and he is unaware of it. Were he to eat a piece of bread which is hard, it would descend into the intestines and scratch them; but the Holy One, blessed be He, created a fountain in the middle of the throat, which enables the bread to move down safely." "If the bladder is pricked by only a needle, all the air in it comes out; but man is made with numerous orifices, and yet the breath in him does not come out."

How unlike the work of man is the handiwork of God! The best of man's work has the mark of his imperfection, but what the Lord has wrought is beyond criticism. "When a human being builds a palace, people often come and criticize. If the pillars were taller, they say, if the roof were only higher, it would be better! But has man ever come and said, If I had three eyes or three hands or three legs, if I walked on my head or my head were turned backward, I should have preferred it? ... The Holy One ... decided upon every limb which you have and set it in its proper place."[8]

MIRACLES

The conception of the universe as the offspring of a plan, as the perfect embodiment of God's design, implied a certain order in its actions. A universe that behaved capriciously would reflect adversely on the plan by which it was fashioned. Thus the rabbis were moved to affirm uninterrupted regularity as one of the characteristics of life in the universe. This did not rule out miracles for them, however. According to one interpretation miracles were provided for at the very time when God brought the universe into being. These events seem deflections from the norm to us, but they are not breaks in the plan which actually made room for them. As the Midrash put it: "At the creation God made a condition with the sea that it should be divided to permit the children of Israel to pass, with the sun and the moon to stand still at the bidding of Joshua, with the ravens to feed Elijah ..."

The age of miracles was not altogether past, however. Some of the leading Talmudists were described as miracle workers. Such stories were associated especially with Rabbi Pinhas ben Yair and Rabbi Haninah ben Dosa. For those who had the sensitivity to see, moreover, there were miracles transpiring daily throughout creation. "Greater is the miracle that occurs when a sick person escapes from perilous disease than that which happened when Hananiah, Mishael, and Azariah escaped from the fiery furnace." And the tortuous manner in which a family in distress manages to eke out an existence is as great a miracle as the parting of the Red Sea for the Israelites.[9]

THE PURPOSE OF HUMAN EXISTENCE

What is the purpose of human life? Why did God bring man upon the arena of existence? It is that he might glorify his Maker through the cultivation of virtue and the continued perfection of his life. The Talmud abounds in discussions as to what is meant by the perfection of life. In the fullest elaboration of their thinking we are offered a vast body of ideals and rules of action by which a person would please his Maker and thus justify his own existence. The principal demand is ethical—to act with compassion and loving-kindness towards God's creatures. Thus Rabbi Akiba pointed to the golden rule as the most comprehensive teaching of the Torah. "This is the most fundamental principle enunciated in the Torah," he taught, "'Love thy neighbor as thyself'" (Lev. 19:18). Ben Azzai made the Torah's fundamental teaching not the golden rule but the doctrine on which it is ultimately based—that man is made in the divine image: "This is the book of the generations of man ... in the likeness of God made He him" (Gen. 5:1).[10]

The Talmudists saw, however, that the anchor on which all the elements of the good life rest, is the recognition of God's sovereignty. It is the reverence for God that ultimately inspires the attitudes and the actions that spell ethical living.

This conception of the relationship between belief in God and the moral life is conveyed in a number of Talmudic discussions. There is the well-known homily by Rabbi Simlai: "Six hundred and thirteen commandments were addressed to Moses—three hundred and sixty-five prohibitions corresponding to the days of the solar year, and two hundred and forty-eight positive commandments corresponding to the number of limbs in the human body. David came and reduced them to eleven principles, which are listed in Psalm 15. Isaiah came and reduced them to six as is said, 'He that walketh righteously and speaketh uprightly, he that despiseth the gain of oppression, that shaketh his hands from holding bribes, that stoppeth his ears from hearing of blood, and shutteth his eyes from looking upon evil' (Is. 33:15). Micah came and reduced them to three, as it is written, 'What does the Lord require of thee, but to do justly, to love mercy and to walk humbly with thy God' (Micah 6:8). Isaiah subsequently reduced them to two, as it is said, 'Thus saith the Lord, keep ye justice and do righteousness' (Is. 56:1). Lastly came Habakkuk and reduced them to one, as it is said, 'The righteous shall live by his faith'" (Hab. 2:4).[11]

GOD AS THE SOURCE OF MORALITY

Because faith in God is the source of the moral life, the rabbis regarded a morality that is not rooted in piety as precariously insecure. And while they placed the love of man at the climax of human virtue, they summoned people to cultivate the love of God as the source from which all other virtues flow.

This is taught by Rabbi Reuben who had been asked to define the most reprehensible act a man may be guilty of. His answer was that it is the denial of God's existence. "For no man violates the commandments, 'Thou shalt not murder', 'Thou shalt not steal', till he has renounced his faith in God."[12]

The same doctrine is conveyed in the famous homily by Raba. As the Talmud relates it, "Raba said: when a person is brought for judgment on Judgment Day he is asked 'Did you do your business honestly, did you set aside time for the study of Torah, did you raise a family, did you maintain our faith in the Messianic redemption, did you pursue wisdom, did you attain to the level of being able to reason inferentially from one proposition to another?' All this will suffice provided he be a God-fearing man, too, for the fear of God is the treasury in which all else is stored. If he be not a God-fearing man, the other virtues will prove insufficient."[13]

THE TOOLS OF HUMAN PERFECTION

The acquisition of a virtuous character and the attainment of life's perfection do not come easily to a man. He must work for them hard and persistently throughout the years, and his gains, such as they are, will always be partial and relative. But God has given man the tools with which he is to make his quest a profitable enterprise. Into his very nature God has poured certain drives which spur him on and guide him on his way. There is the impulse to look after one's self. This is sometimes called the evil impulse, because when carried beyond its legitimate limits the preoccupation with one's self becomes a destructive force in human life. But in its essential character this impulse is no more evil than anything else which the Lord has made. Balancing this impulse, moreover, is the drive to goodness, the *yezer tob*, which spurs us on to acts of self-denial in furtherance of every noble endeavor. In present circumstances the so-called evil impulse dominates life, but as men mature in their development the good impulse gains ascendency and the proper balance is achieved between those two basic drives of our natures. The Talmudists pronounce their judgment on the two impulses in a comment on Genesis 1:31: "And God saw everything which He had made and behold it was very good." "Very good," say the rabbis, applies to those two impulses. "But," it is asked, "is the evil impulse very good?" And the answer is given that it is. For "were it not for that impulse, a man would not build a house, marry a wife, beget children or conduct business affairs."[14]

The person in whom the drive for self has been integrated in a sound pattern of character has made of the so-called evil impulse also a tool of goodness. The Talmud makes this clear in the comment on Deut. 6:5: "Thou shalt love the Lord thy God with all thy heart." "Thy heart" is taken as meaning "with the two impulses—the good and the evil."[15]

65

THE TORAH AND THE COMMANDMENTS

God moreover did not thrust man into the world to grope entirely on his own for the right course he must pursue in life. He has given man a chart by which he can steer himself. This chart is contained in the Torah and the commandments.

The Talmudists believed firmly that God revealed Himself to man, that having formed human life, He is also concerned with guiding it toward the knowledge of virtue and truth in which man finds his true happiness. Not every person is ready to receive the divine revelation. But there are some who are ready, and to them God reveals Himself. Those chosen few are, however, chosen not for their own edification, but that they might become His prophets, the instruments for disseminating the fruits of that revelation among all mankind.

The most important manifestation of prophecy was in Israel, but not exclusively so. The rabbis saw the evidence of prophetic inspiration in the lives of men outside the Jewish people. Thus they declared: "Seven prophets prophesied for the pagans:[16] Balaam and his father, and Job and his four friends." But prophecy, in its highest expression, appeared in Israel solely.

The most important permanent fruit of prophecy in Israel were the various books that make up the Holy Scriptures, commencing with the Pentateuch which is traced back to the authorship of Moses. In the words of the famous statement of the Talmud: "Who wrote the Scriptures? Moses wrote his own book and the parables of Balaam (Nu. 23, 24) and Job; Joshua wrote the book which bears his name and the last eight verses of the Pentateuch; Samuel wrote the book which bears his name and the Book of Judges and Ruth; David wrote the Book of Psalms ... Jeremiah wrote the book which bears his name, the Book of Kings and Lamentations; Hezekiah and his colleagues wrote Isaiah, Proverbs, the Song of Songs, and Ecclesiastes; the men of the Great Assembly wrote Ezekiel, the Twelve minor Prophets, Daniel and the Scroll of Esther. Ezra wrote the book that bears his name and the genealogies of the Book of Chronicles up to his own time ..."

The degree of divine inspiration bestowed on individual prophets varied greatly. Moses was supreme among them, and the quality of his inspiration was surpassed by none. But even among the other prophets there were individual differences. Isaiah, for instance, was held superior to the others. Thus the Midrash suggests: "The Holy Spirit descends on the prophets in degrees. Some prophesied to the extent of one book, others of two books. Beeri only prophesied two verses, which, being insufficient for an independent book, were included in Isaiah."[17]

The Divine plan could not, however, fulfill itself through the individual prophets. It was essential that the prophets be given a particular society which would be most responsive to their call and that would be prepared to dedicate its common life to the implementation of their ideals.

For that special duty God chose Israel. A Talmudic homily relates how God sought out the society that was best prepared to be the custodian of the Torah. "When the All-present revealed Himself to give the Torah to Israel, not to them alone did He manifest Himself, but to all the nations. He first went to the sons of Esau, and said to them, 'Will you accept the Torah?' They asked what was written in it and God told them: 'Thou shalt not murder.' They replied, 'Sovereign of the Universe! The very nature of our ancestors was bloodshed …' He then went to the sons of Ammon and Moab and said to them, 'Will you accept the Torah?' They asked what was written in it and He replied, 'Thou shalt not commit adultery!' They said to Him, 'Sovereign of the Universe! The very existence of this people is rooted in unchastity.' He went and found the children of Ishmael and said to them, Will you accept the Torah?' They asked what was written in it and He replied: 'Thou shalt not steal.' They said unto Him, 'Sovereign of the Universe! The very life of our ancestors depended upon robbery …' There was not a single nation to whom He did not go and offer the Torah …" The selection of Israel, in other words, was not arbitrary. God selected Israel "because all the peoples repudiated the Torah and refused to receive it; but Israel agreed and chose the Holy One, blessed be He, and His Torah." Israel was the chosen people in a double sense. Israel had chosen God even as God had chosen Israel. Israel's function in history, then, was to serve as a witness to the truths of the Torah. For the Torah of which Israel was the custodian was ultimately intended for all mankind. It is for this reason that the original promulgation of the Torah took place in the desert, a no-man's land, rather than in the land of Israel. This was to suggest that its treasures were not meant to belong to any particular people exclusively; the Torah was God's message through Israel to all humanity.[18]

In projecting the goal of sharing the Torah with the rest of mankind, the rabbis did not call for the conversion of the rest of the world to Judaism. They distinguished between a universal element in their faith which all men must adopt and a more particular element which applied to the more specific facts of the Jewish group itself. This universal element of Judaism to which all men were summoned could be integrated with any culture and with whatever formal expression had developed in the religious life of a people. Its provisions are known as the "Seven Noahite laws" and they include the practices of equity in human relations, the prohibition of blaspheming God's name, the prohibition of idolatry, sexual unchastity, bloodshed, robbery, and cruelty to animals, such as tearing a limb from the animal when it is still alive.[19]

Proselytes were of course accepted in Judaism, when they proved their sincere desire to become part of Israel and to share in its destiny. But that, the rabbis made it clear, was not a prerequisite for earning divine approval. "A pagan," declared Rabbi Meir, "who studies the Torah and practices it is the equal of a high priest in Israel."[20] Rabbi Meir clearly refers to a pagan who practices the universal principles of religion and morality as embodied in the so-called "Seven Noahite laws". If he practiced the Torah in its entirety he would no longer be a pagan.

The Talmud makes the study of Torah a cardinal virtue in Judaism and summons all men to engage in it. "Whoever labors in the Torah for its own sake," declares the Mishnah, "merits many things; and not only so, but all creation is vindicated through him. He may be acclaimed as friend, beloved, a lover of the All-Present, a lover of mankind. It clothes him in meekness and reverence; it enables him to become just, pious, upright, and faithful; it keeps him far from sin, and brings him near to virtue. Through him the world enjoys counsel and sound knowledge, understanding and strength. ... It also gives him sovereignty and dominion and discerning judgment. The secrets of the Torah are revealed to him. He is made like a never-failing fountain, and like a river that flows on with ever sustained vigor. He becomes modest, long-suffering, and forgiving of insults; and it magnifies and exalts him above all things."[21]

Life's highest goal which is attainable by man must be sought by living according to the teachings of the Torah. The study of Torah must therefore be the great preoccupation of mankind. "The ignorant man cannot be pious," as Hillel puts it, and ignorance here refers clearly to ignorance of Torah. In poverty, in wealth, in youth and old age, a person must ever give himself to the mastery of Torah. It is the only sure compass by which he can guide himself amid the turbulence and uncertainty of life about him.[22]

The rabbis saw the educational service of the Torah reinforced by the disciplines which are enjoined by it. Some of the commandments enjoined in the Torah are clearly ends in themselves. Thus the many prescriptions in civil and criminal law aim at creating a just order of human relations. Many of those commandments, however, were enjoined for pedagogic reasons—to teach certain truths through the more dramatic affirmation of action. They were meant to teach as reminders of vital truths, like the *mezuzah*, the receptacle attached to the door post of the house and the phylacteries worn on arm and head during prayer. Both contain parchments on which is written the text of the most important injunction of Scripture—to "love the Lord thy God with all thy heart, and all thy soul and all thy might" (Deut. 6:5). By the continued exposure to these symbols man was to be reminded vividly of his relationship to God. As the rabbis put it: "Whoever has phylacteries on his head and arm, the fringe on his garment and the mezuzah on his door may be presumed to be safe from committing sin."[23]

The performance of the commandments was seen as serving man in a deeper sense. It gave him the opportunity to do something concrete in implementation of his love of God, thereby ennobling his own character. As one rabbinic comment expressed it: "The commandments were only given for the purpose of refining human beings; what, for example, does it matter to the Holy One, blessed be He, whether an animal's neck is cut in the front or the rear (as prescribed in the dietary laws)! But the ordinances He gave us have as their purpose the purification of human beings."[24]

FREEDOM AND ITS CONSEQUENCES

The conquest of human lives for the truths of the Torah is a painfully slow process. Even Israel, who carries the special responsibility of being the servant of the Lord in the propagation of the Torah, frequently falls so far short of its ideal. And sin, the defiance of God's word, seems to be the all-pervasive failing among human beings. No doubt, God could have made men without the capacity to err. That. however, would have destroyed human freedom. Instead. God has made man a free agent, which involves the uncoerced exercise of the will in any direction, regardless of its moral consequences. As an oft-quoted Talmudic maxim: "All is in the hands of Heaven except the fear of Heaven."[25] God, in other words, is master of the Universe, but He is not master over man's moral decisions, which he must learn to make himself. But man is not left to his own initiative exclusively. God aids him in learning to exercise his freedom in ever wiser decisions. For whenever men defy the truths of the Torah and build patterns of personal and group life in violation of its teachings, God passes judgment upon them; and the discipline of suffering reinforces the native appeal of truth itself in leading man to repentance. It is in this spirit that the Midrash interprets verse in Psalm 23: "Thy rod and Thy staff they comfort me"; rod is applied to suffering while staff is applied to the Torah. Suffering is therefore not an evil to be avoided but an opportunity that points to a better life. "Whoever rejoices in the sufferings that come upon him in this life brings salvation to the world."[26]

The Talmudists did not advise people to seek suffering. One of them put it quite bluntly: "I desire neither the suffering nor the rewards which it brings in its train." But when suffering comes, we are to see it as the prodding of God who is displeased with us for having committed sin, and who is bestowing upon us the favor of pushing us toward new religious and moral growth. In the words of the Talmud: "Should a man see suffering come upon him, let him scrutinize his actions; as it is said, 'Let us search and try our ways, and return unto the Lord' (Lament. 3:40). If he has scrutinized his actions without discovering the cause, let him attribute them to the neglect of Torah, as it is said, 'Happy is the man whom Thou chastenest, and teachest out of Thy Law' (Ps. 94:12). If he attributed them to neglect of Torah without finding any justification, it is certain that his chastenings are chastenings of love; as it is said, 'For whom the Lord loveth He correcteth'" (Prov. 3:12).[27]

The so-called suffering inflicted because of "love" is the highest kind of suffering. For it comes not to expiate for some wrong done but to disturb life's stagnation and to initiate a new spiritual advance. It is the irritant that stimulates spiritual progress. It was no doubt because he viewed life from this perspective that one rabbi paid tribute to God for the very sufferings He had inflicted on Israel: "Because God loved Israel He multiplied sufferings for him." For through such sufferings Israel would achieve a new vitality in its spiritual life. "As the olive does not give of its precious oil except under pressure, so Israel does not bring forth its highest virtues except through adversity."[28]

In their trials no less than in their triumphs, therefore, God is guiding mankind toward their destiny. But its fulfillment is a long process toward which men climb slowly in their varied vicissitudes of history. When the theme of history reached its climax, the Talmudists were confident there would be ushered in a state of unusual human perfection. Then men will become completely reconciled with God and surrender unreservedly in loving obedience to His will. Oppression and hatred will then disappear and a new order of righteousness and love will be established in the world. It will involve the full realization of the hopes of the prophets and the fulfillment of Israel's mission in history. And it is to be brought about through a human instrument, the Messianic deliverer.

THE MESSIANIC HOPE

There is a wealth of varied details with which different rabbis surrounded the belief in the Messiah. But certain essential features stand out. The term Messiah means anointed, an allusion to the installation ceremony of kings and priests in their respective offices. But the Davidic dynasty carried so many fond associations among the Jewish people and recalled a glorious period in Jewish history, it was generally assumed that "The anointed" would be a scion of the house of David. His arrival will take place after a great suffering will have regenerated the hearts of men; they will have to suffer the pangs that are attendant upon every new birth, pangs that are therefore designated the "travail of the Messiah". The human regeneration which is to usher in the Messianic fulfillment, moreover, will not be complete. There will be men who will hold on defiantly to the error of their life and endeavor to impede the dawn of the new day. And those men will have to be vanquished in a bloody contest of arms.[29]

As for the time when this consummation was to take place, it was generally held to depend on the degree of progress men will have achieved in their development. This is well illustrated in the well-known Talmudic parable. "Rabbi Joshua ben Levi met Elijah standing at the entrance of Rabbi Simeon ben Johai's tomb. ... He then said to him, 'When will the Messiah come?' 'Go and ask him' was the reply. 'Where is he sitting?'—'At the entrance of the city.' ... So he went to him and greeted him, saying, 'Peace be upon thee, Master and Teacher.' 'Peace be upon thee, O son of Levi,' he replied. 'When will thou come, Master?' asked he. 'Today' was his answer." When the Messiah failed to appear that day, a deeply disappointed Joshua returned to Elijah with the complaint: "He spoke falsely to me, stating that he would come today, but has not!" Elijah then enlightened him that the Messiah had really quoted Scripture (Ps. 95:7): "Today, if ye hearken to His voice."[30]

SOCIAL ETHICS IN THE TALMUD

THE UNITY OF MANKIND

The Talmudic conception of mankind is that of a unity, deriving its character from a common origin and a common destiny. The basic elements of this doctrine are already enunciated in the Bible which traces the origins of the human race to a single person who is formed by God in His own image. It is in the Talmud, however, that this doctrine reaches its fullest maturity. "Why did the Creator form all life from a single ancestor?" inquired the Talmud, and the reply is, "that the families of mankind shall not lord one over the other with the claim of being sprung from superior stock ... that all men, saints and sinners alike, may recognize their common kinship in the collective human family."[1]

Human behavior may be infinitely varied, but human nature which underlies it, is essentially the same. Man is a creature of earth and at the same time a child of God, infused with the divine spirit. Appraised in moral categories, all people are endowed with the tendency to see in their own persons the ultimate ends of their being and the tendency to seek transcendent ends toward which their own persons are but contributing instruments. Out of these two tendencies flow good and evil, which thus reside, in varying measure, to be sure, in every individual as part of his indigenous equipment for life. If you but probe sufficiently, one Talmudic maxim advises, you will discover that "even the greatest of sinners" abound in good deeds as a pomegranate abounds in seeds. On the other hand, the greatest of saints have their share of moral imperfection.[2] All human beings are, so to say, cut from the same cloth and there are no absolute distinctions between them.

THE UNIQUENESS AND SANCTITY OF HUMAN LIFE

This doctrine of equality does not assert that individuals duplicate one another. "A man," the Talmud explains, "strikes many coins from one die and they are all alike. The Holy One, blessed be He, however, strikes every person from the die of the first man, but no one resembles another." Their uniqueness is mental as well as physical, and they all have a special function to fulfill in the realization of the cosmic purpose. A person thus has a right to feel that "the universe was created for his sake," for he has a unique role to play in it, so that the cosmic scheme will be incomplete without him.

The specific role that one's particular faculties enable him to play is immaterial. Humble or exalted, all roles are equally invaluable to the fulfillments of history. In the words of a Talmudic illustration, "I am a creature of God and my neighbor is also His creature; my work is in the city and his is in the field; I rise early to my work and he rises early to his. As he cannot excel in my work, so I cannot excel in his work. But you may be tempted to say, 'I do great things and he small things!' We have learned that it matters not whether one does much or little, if only he directs his heart to serve the divine purpose."[3]

Deriving from this conception of man's place in the universe is the sense of the supreme sanctity of all human life. "He who destroys one person has dealt a blow at the entire universe, and similarly, he who makes life livable for one person has sustained the whole world." All law, civil and religious, has as its purpose the promotion of human life, and when it ceases to serve that end it becomes obsolete and is to be superseded. To quote a good Talmudic maxim, "The Sabbath is made for man and not man for the Sabbath"; and what was true for the Sabbath applied likewise to all other law. It is greater to serve one's fellow-man, one Talmudist expounded, than to preoccupy oneself with divine communion.[4]

The sanctity of life was intrinsic to the individual person and was not a derivative of national origin, religious affiliation, or social status. As one Talmudist generalized: "Heaven and earth I call to witness, whether it be an Israelite or pagan, man or woman, slave or maidservant, according to the work of every human being doth the Holy Spirit rest upon him." Non-Jews residing in Jewish communities were to share in all the beneficences which the Jewish community held out to its own members. Jews were ordained to sustain their needy, to visit their sick, and to bury their dead. As the rabbis put it: "We are obligated to feed non-Jews residing among us even as we feed Jews; we are obligated to visit their sick even as we visit the Jewish sick; we are obligated to attend to the burial of their dead, even as we attend to the burial of Jewish dead." The rabbis base their demand on the ground that these are "the ways of peace."[5]

Nor was a person's worth a derivative of his status, whether political, social or cultural. In the sight of God the humble citizen is the equal of the person who occupies the highest office. The Talmud did not outlaw slavery which was an integral part of ancient economy, but it sought to limit its degrading aspects. Already Biblical law had declared a Hebrew slave free after a seven year period of service. Talmudic legislation continued to extend the solicitude on behalf of the slave's welfare. The slave was to live at the same level of comfort as was enjoyed by his master. "Do not eat fine bread and give black bread to your servant, do not sleep on cushions and have him sleep on straw." So exacting was the Talmud in its defense of the slave's dignity that it became a proverbial expression, "Whosoever buys a Hebrew slave, buys a master unto himself." Indeed, the Hebrew slave was really a workman who had temporarily sold his services but whose dignity and rights remained intact. And the Talmud condemned the man who was willing to accept personal bondage as a solution to his economic problem; for man was meant to serve only God and to recognize no other master beside Him.[6]

But the Talmud includes equally telling expressions of solicitude on behalf of the pagan slave. He was not to be exposed to ridicule or humiliation. One Talmudist shared his meat and wine with his slave, explaining: "Did not He that made me in the womb make him also?" It was an old principle which the Pharisees had established that "slaves, unlike the ox or the ass, are human beings with minds and wills of their own."[7]

The Talmud speaks repeatedly of the dignity of free labor. Creative labor, no matter how humble, is always honorable and is a form of divine worship, for it contributes to the maintenance and development of civilization. "Flay dead cattle on a highway," runs a Talmudic proverb, "and say not 'I am a priest, I am a great man and it is beneath my dignity.'" One of the responsibilities which every parent owes his son is to teach him a trade. The Talmudists, themselves, because their academic work was a labor of love which offered no remuneration, pursued various handicrafts as well as farming and commerce to earn a livelihood. Among them were shoemakers, tailors, bakers, woodcutters, a night watchman and even a grave digger.[8]

Even he who had endangered social security in the commission of crime has not forfeited his inherent worth as a person. The Talmud ordained with great emphasis that every person charged with the violation of some law be given a fair trial, and before the law, all were to be scrupulously equal, whether a king or a pauper. One of two litigants was not to appear in court in expensive robes when the other came in tatters, lest there be a swaying of the juror-judges.[9]

Particularly in criminal cases did the Talmud seek to protect the accused against a miscarriage of justice. Circumstantial evidence, however convincing, was not acceptable. At least one of the judges was to act as the counsel of defense. The juror-judges could reverse a vote from guilty to not guilty, but not vice versa. The younger members of the court were first to announce their vote, so as not to be influenced by the actions of their seniors. Whereas in civil cases a majority of one was sufficient to establish guilt, in criminal cases a majority of two was required.

Even when he was found guilty, he had not lost his link to the human brotherhood. The larger ends of safeguarding the community may require his extermination, but whatever punishment is inflicted upon him must be humanized by a persistent love and not brutalized by vengeance. Certain Talmudists advocated the abolition of capital punishment, and it was agreed that any court that inflicts capital punishment once in seven years had exhibited brutality. The execution even of the most violent criminal is a cosmic tragedy. For he, too, was formed in the divine image and had been endowed with infinite possibilities for good.[10]

In the hierarchy of Jewish values the knowledge and practice of the Torah represented the apex, but the master of the Torah was not to hold himself aloof from or superior to other men. He was to be "modest, humble … to make himself beloved of men, to be gracious in his relations even with subordinates … to judge man according to his deeds." To show pride in one's learning is to become "like the carcass of a dead beast from which all men turn away in disgust." The true master of Torah will be inspired by a greater learning and piety not to aggrandize himself over others or to detach himself from the common people and cultivate his virtues in the privacy of his own home, but to teach and lead the common people to a nobler way of life. He who has insights that can broaden the horizons of his neighbor's life and does not communicate them is robbing his neighbor of his due. The gifts of the spirit, like the gifts of substance, are a trust to be shared with others.[11]

CONSENT AND THE MAJESTY OF THE LAW

Throughout Talmudic times the Jews lived under the domination of foreign imperialisms; in Palestine under the Romans and in Babylonia under the Parthians and neo-Persians.

Whether a free Jewish commonwealth would have developed a democratic representative government, we do not know. But within the framework of the limited autonomy which the Jews enjoyed, they did develop certain democratic institutions. The most important instrument of Jewish autonomy was Jewish civil and religious law, and the Talmud developed the theory that the ultimate sanction of all law is the consent of the people who are to be governed by it. For the Talmud, of course, all authority, including the authority behind the makers and interpreters of law flowed from the divine source which manifests itself in every form of human leadership. But man is endowed with free will and his unrestrained conscience must give its assent to every legal institution that is to have moral claims over him. Judges and legislators must not enact decrees unless a majority of the people find it possible to conform to them. Any decree which is resisted by a popular majority has, *ipso facto*, lost its validity and been rendered obsolete. Indeed, the Talmud even traced the authority of the Bible itself not so much to its divine source as to the consent of the people who fully agreed to live by it.[12]

Social stability frequently calls for disciplined behavior; and in the field of social and religious conduct, the Talmud called upon the individuals to conform to the majority decisions of the duly constituted authorities who interpreted Jewish law. In the field of opinion, however, the individual remained essentially free to believe and speak in accordance with the dictates of his own conscience. Indeed, there has never been formulated an official creed in Israel as a criterion of loyalty to the mandates of Jewish life. And even in law, the minority could continue defending its position in the hope that the majority might eventually be moved to reconsider its judgment. As the Talmudists put it, majorities and minorities are equally "the words of the living God"; they both represent aspects of truth, and are equally precious. The Talmudists themselves preserved all dissident opinions which developed in their discussions and even recorded them side by side with the majority opinions which became authoritative law.

The Talmudists developed a system of democratically constituted town councils which were charged with the administration of local municipalities. All those residing in a community for a year or over enjoyed the right to participate in the election of the seven town councillors. The functions of these town councils were far-reaching, including the supervision of economic, religious, educational and philanthropic activities of the people. On important issues, town meetings were held in which the will of the people could be ascertained more directly. Certain local officials were of course appointed by the head of the Jewish community, the patriarch in Palestine, and the exilarch in Babylonia. But the most important requirement in all such appointments was that they meet with the public approval. In the words of the Talmud, "We must not appoint a leader over the community without first consulting them, as it is said, 'See, the Lord hath called by name Bezalel, the son of Uri' (Exodus 35:30). The Holy One, blessed be He, asked Moses, 'Is Bezalel acceptable to you?' He replied, 'Sovereign of the universe, if he is acceptable to Thee, how much more so to me!' God said to him, 'Nevertheless go and consult the people ...'"[13]

SOCIAL WELFARE AND PERSONAL FREEDOM

The social process frequently brings individuals into a position where they exercise power over the lives of others. In the social theory of Talmudic Judaism, it then becomes the task of the community to develop such instruments of social control as will rationalize that power with moderation and justice. The Talmudists declared individual property rights as subject to their consistency with the public welfare. When it is to serve the public interest, these rights may be modified or suspended altogether. Basing its action on this principle, Talmudic legislation regulated wages and hours of labor, commodity prices and rates of profit. They held it was similarly the task of the community to provide other facilities for promoting the public welfare, such as public baths, competent medical services, and adequate educational facilities for all, at least on an elementary level.[14]

The poor had a claim upon the community for support in proportion to their accustomed standard of living. The more affluent individuals were to share their possessions with them, as members of a family circle were obligated to share with their own kin. To place the administration of poor relief on a more efficient and respectable basis, it was eventually institutionalized. Begging from door to door was discouraged. Indigent townsmen were given a weekly allowance for food and clothing. Transients received their allowance daily. Ready food was also kept available to cope with immediate needs. For the poor traveler and the homeless, public inns were frequently built on the high roads. All these facilities were maintained from the proceeds of a general tax to which all residents of a community contributed.[15]

Perhaps the most interesting form of poor relief, from a modern standpoint, is a public works project for the assistance of the unemployed, the details of which have been preserved by Josephus but which was instituted in Talmudic times: "So when the people saw that the workmen were unemployed who were above 18,000 and that they, receiving no wages, were in want … so they persuaded him (King Agrippa) to rebuild the eastern cloisters; … he denied the petitioners their request in the matter; but he did not obstruct them when they desired the city might be paved with white stone …"[16]

DEMOCRACY AND FAMILY LIFE

The same concern for the values of humanitarianism and democracy appears in the Talmudic legislation bearing on the various aspects of family life. The Talmud does not regard the individual man as a self-sufficient personality. He is completed through matrimony. "The unmarried person lives without joy, without blessing and without good. He is not a man in the full sense of the term; as it is said (Genesis 5:2), 'male and female created He them, and blessed them and called their name man.'"

Happiness in married life involves many compromises, but these must be assumed in freedom. They should not be imposed through constraint from any external source. In the words of the Babylonian teacher Rab, "A man is forbidden to give his minor daughter in marriage without her consent. He must wait until she grows up and says 'I wish to marry so and so.'" If he did give her in marriage as a minor, she could protest the marriage on reaching maturity, and have it annulled without divorce. The man's choice, too, should be voluntary and an expression of considered choice. "A man should not marry a woman without knowing her lest he subsequently discover blemishes in her and come to hate her."[17]

As the more dominant partner in the family circle, the husband was exhorted to treat his wife with tenderness and sympathetic understanding. "Whoever loves his wife as himself and honors her more than himself … to him may be applied the verse, 'Thou shalt know that thy tent is in peace.'" Before the children, father and mother were equals. They were both to be accorded the very same devotion and respect.[18]

The Talmud regards divorce as the greatest of all domestic tragedies. "Whoever divorces the wife of his youth, even the altar sheds tears on her behalf, as it is written, 'And this again ye do; ye cover the altar of the Lord with tears … because the Lord hath been witness between thee and the wife of thy youth, against whom thou hast dealt treacherously.'" There are occasions, however, when husband and wife cannot harmonize their natures and irreconcilable differences develop between them. The Talmud then sanctions divorce, as preferable to a life of continuing bitterness and distress.

Divorce could be achieved upon the considered request of either party. Theoretically, it was always the husband who severed the marriage ties and not the wife. But the wife could sue for divorce and, if the request seemed warranted, the court forced the unwilling husband to divorce her. Among the circumstances warranting such action by the court, the Talmud lists the husband's impotence, failure of proper support, denial of conjugal rights, contraction of a loathsome illness, or engaging in a repugnant occupation. The divorced woman was protected by the Ketubah or marriage contract, which provided a financial settlement for her maintenance.[19]

For the Talmudists, children are the noblest fulfillment of married life. For it is man's elemental duty to the continuity of life to bring children into the world and to raise them properly. Nevertheless, where conception was likely to prove dangerous to the mother, birth control was recommended. In the words of the Talmud, "Three types of women should employ an absorbent to prevent conception: a minor, a pregnant woman, and a nursing mother; a minor lest pregnancy prove fatal, a pregnant woman lest she have an abortion, and a nursing mother because of the danger to her young infant."[20]

The Talmud offers detailed advice on how to bring up children. Parents must treat all children equally and avoid any display of favoritism between them, which can only lead to jealousy and family discord.

Parents must not over-indulge their children, which is the surest road to character depravity. Thus the Talmudists blame the depraved character of Absalom who led a revolt against King David, his father, to his pampered youth. But excessive severity is no less harmful. The Talmud cites the case of a child who committed suicide after some petty misdeed because he was in such mortal fear of his father.

The Talmud ordains a profound respect which children owe to their parents. Even he who begs from door to door is committed to provide for the sustenance of his needy parents. But the proper respect due parents is not merely a matter of material help. The intangibles of tenderness and consideration are equally important. To cite a Talmudic illustration, "There was a person who fed his father on fat poultry. Once his father asked him, 'My son, where do you get all this?' To which he replied, 'Old man, eat and be quiet, for dogs eat and are quiet.' Though he fed his father fat poultry, such a person will inherit Gehinnom."[21]

EDUCATION AND THE COMMON MAN

Perhaps the most significant triumph for democracy in Talmudic Judaism was the development of a system of free, universal education. The Jewish school system began with higher rather than elementary education. The most important institution of higher education was the Sanhedrin itself and the hierarchy of various lower courts which functioned under its supervision. Their deliberations were made accessible to advanced students who were preparing themselves for ordination; and they were even permitted to participate in the discussions. Witnessing the conflicts of personalities, the play of minds, and the manipulation of dialectic by which the Torah supplementation was evolved, represented a vivid and unforgettable educational experience. In addition the leaders of Pharisaic and rabbinic Judaism conducted a formal instruction in their own schools. Some of these schools were particularly famous. The schools of Shammai and Hillel were continued even after their founders were gone. Akiba's school which was finally conducted at B'nai Brak is said to have had an enrollment of 12,000 students, like a modern metropolitan university.

In early times, these schools charged tuition fees which were payable upon admission to each lecture. And many made great sacrifices to attend, frequently working their way through school. This is vividly illustrated in the famous story of Hillel's struggle for an education. Hillel spent half of his daily earnings for admission to the lectures in the academy of Shemaya and Abtalyon. One winter day, being out of work, he could not pay the necessary admission charge, and the doorkeeper refused to admit him. Deter. mined not to miss the session, he climbed up the roof and listened to the discussion through the skylight. On the following morning the room was darker than usual; and looking up at the skylight, they saw the figure of a human body. Hillel had been snowed under. Fortunately the discovery had been made in time, and Hillel was saved. This admission fee was abolished after the destruction of the Temple and higher education became wholly free. In addition, lectures were offered in the evening which facilitated attendance for those who had to work for a livelihood during the day.[22]

Elementary education was originally left to the home, but in time this too was institutionalized. As the Talmud relates it: "Were it not for Joshua ben Gemala (high priest who was in office in the latter part of the first century), the Torah would have been forgotten in Israel." In antiquity every father taught his own child. Those who were without fathers to teach them were thus left without education. Later on, schools were established in Jerusalem to which the children were to be sent from all over the country. But these too were inadequate. Thereupon they established regional schools to which youths of 16 or 17 were admitted. But it was soon apparent that adolescents could not first begin to subject themselves to school discipline. "Rabbi Joshua then instituted schools in each province and town and children were enrolled at the age of six or seven." Classes were generally conducted in the synagogue buildings, though they were frequently transferred to the outdoors. There were, according to the Talmud, three hundred and ninety-four schools in Jerusalem before its destruction by the Romans in 70 C.E. The curriculum concentrated on Biblical literature, Midrash and, later on, also on the Mishnah.

The rabbis were equally devoted to educating the general public. Their formal lectures in the schools were generally open to lay auditors. In addition they utilized the synagogue service which brought out large numbers, as an opportunity for educational work. The liturgy itself, which was eventually recited thrice daily by every Jew, was an affirmation of the fundamental beliefs of Judaism. Readings from the Torah, with appropriate elucidations in the Aramaic vernacular, had been made an integral part of the synagogue ritual ever since the days of the Sopherim. Four times weekly, Saturday morning and afternoon, Monday and Thursday, as well as on all feasts and holidays, and on the new moon, the Jewish laity thus listened to Scripture lessons.

Under the inspiration of the Synagogue, smaller groups of people formed into individual study circles meeting at convenient hours on weekdays or the Sabbath for the study of Scriptures or some other branch of Jewish tradition. This was later enhanced with the introduction of the popular sermon Friday evening and Saturday morning, and there were special sermons before each holiday.

Some of the rabbis were not particularly gifted with eloquence, and it therefore became customary for an additional functionary to attach himself to the rabbi, the orator-commentator. In academy and synagogue alike, such a rabbi would first communicate his message to the commentator who then made this the theme of his oration before the public. The synagogues in every community, in addition to providing for religious worship also functioned as popular universities diffusing the knowledge of the Torah among the common people.[23]

THE NATION AND THE WORLD COMMUNITY

The sanctity of human life implied for the Talmudists a similar concern for the national community. For each society, too, makes its unique contribution to the fulfillments of history. The Talmudists speak of Israel as being particularly creative in the field of religion, whereas other peoples achieved comparable distinction in other fields—in the arts and sciences. There were some who spoke with admiration of Roman law, of the Roman system of public markets, bridges and baths. The collective welfare of all humanity is contingent upon the welfare of every individual people, and the sacrificial cult of the second Temple in Jerusalem included, during the Feast of Tabernacles, seventy offerings invoking God's aid for each of the seventy nations of the world.

The aberration of human sin will occasionally drive groups to seek dominion over others. Thus in Talmudic times, the Jews suffered heavily from the oppression of Roman imperialism. The Talmudists decried this oppression and encouraged their people's resistance to it. As we have already noted, they denounced the Jewish tax farmer as a reprobate and robber because he collaborated with the Roman system of extortion and oppression. Deceiving the Roman tax collector they put on a par with deceiving a pirate, for Rome had no moral right to the country which she had occupied by force. The Pharisaic ostracism of the publican, which was but another name for the Jewish tax collector, was not, as has frequently been interpreted, an expression of self-righteousness. It was the reaction of liberty-loving men against those who, for a consideration, were willing to make themselves the partners of an alien imperialism in the plunder and oppression of their own people.

At the same time, the Talmudists guarded against transmuting the temporary historical struggles of their people against various imperialist oppressor-states into enduring hatreds against other nations. The Talmudists spoke with compassion about the vanquished Egyptians who drowned in the Red Sea in a vain pursuit of the fleeing Israelites. Thus they describe God as silencing an angelic chorus which chanted hallelujahs when the Egyptian hosts met their disaster. "My handiwork is perishing in the sea; how dare you sing in rejoicing!"

Even in the face of the tragedy inflicted upon their people by the Romans, the Talmudists sought to avoid hatred. Individual teachers spoke sharply in denunciation of Roman tyranny. But their collective reactions as summarized for instance in the liturgy of that day, is dedicated not to the denunciation of Rome, but to Jewish self-criticism. "It is because of our sins that we have been banished from our land," is the principal motif in the liturgical reaction to the national disaster. And the way of redemption toward which they were taught to strive was moral regeneration in their inner personal and social lives and the interpenetration of the same ideals of a loftier morality among all mankind. In time, the strife of nations, like the strife of individuals, will come to an end in the discovery of their universal interdependence. Israel's cry for justice will be vindicated in a universal fulfillment when the "kingdom of wickedness" shall pass away and all mankind join to form "one fellowship to do the divine will with a perfect heart" (from the liturgy of the New Year, composed by Abba Areka, d. 247).[24]

THE DOCTRINE OF RESPONSIBILITY

But the Talmudic conception of man implied a reciprocal responsibility from individual men and nations to the collective human community. For the fulfillment of the larger organism is dependent upon the integrated functioning of its constituent parts. The unique gifts of energy, substance, or spirit with which an individual is endowed must all be directed to larger human service. As one Talmudist interprets it, the second commandment ordains not alone repose on the seventh day of the week, but also creative labor on the six days. "For is it not written, 'Six days shalt thou do thy work, and on the seventh day shalt thou rest'?" The Talmud denounced asceticism, even when religiously motivated, as sinful, for it withdrew essential creative energies from the tasks of civilizations.

The responsibilities of service rest similarly on every society. And the Talmud called upon the Jews to share with the rest of mankind their achievements in the field where they believed they had distinguished themselves, the field of religion and morality. According to the Midrash, the Torah was originally revealed in the desert and not in the land of Israel, in order to suggest that its teachings were meant for all mankind and not for a particular people exclusively.

Implementing the ideal of its mission, the Judaism of the early Talmudic period proselytized extensively throughout the pagan world. Judaism became, in the words of Professor George Foote Moore, "the first great missionary religion of the Mediterranean world." Because it conceded salvation even to those who were outside its fellowship, Jewish missionaries did not seek only formal conversions; with equal diligence they sought to make what were known to the Romans as *metuentes*, or "God-fearing men," sympathizers of Judaism who, while not conforming to the Jewish ceremonial discipline, would yet order their lives by Jewish ideals of personal and social morality. Through this dissemination of the unique values in Jewish tradition, the Jewish people were to meet their responsibilities to the larger human community of which they recognized themselves to be a part, and to whose service they saw themselves committed by the God who had made them a distinct people in civilization.[25]

PERSONAL MORALITY IN THE TALMUD

The Talmud is concerned with man himself, and not only with the social consequences of his actions. Scattered throughout Talmudic literature, we have therefore a description of the ideal in human character. It is inspired by the religious and moral values which are taught in Talmudic Judaism.

CONFIDENCE IN LIFE

The basic attitude which the Talmudists prized in people is a disposition of confidence in life. Such confidence flows directly from faith in God. For if God's providence extends to all His creatures, then we may be certain that whatever transpires is for the best—at least for the best of creation. A well-known Talmudic maxim reads: "Whatever the Lord does is for the best."

There are occasions when events transpire that we judge injurious to ourselves. In many instances, however, they are really to our advantage, though we may not be aware of it at the time. The Talmud cites an anecdote from the career of Rabbi Akiba which illustrates this truth. While on a journey he sought hospitality in a certain town, but he was turned down, and he had to spend the night in the field. That very night robbers came and plundered the entire town. "He thereupon said to the inhabitants, 'Did I not tell you that whatever the Holy One, blessed be He, does is for the best!'"[1]

The rabbis urged a man to labor diligently in order to provide for himself and his family. "One must not depend on miracles," is a familiar maxim in the Talmud.[2] But once a person assumes his obligations and acts on them he need not be unduly anxious about his livelihood. "A person who has today's bread in his basket and is worried, 'What will I eat tomorrow?'—is a man of little faith," declared Rabbi Eliezer.[3] The Lord stands behind our own endeavors, and as He provides for the raven in the field, He provides for man also. In the words of the rabbis: "He who created each day provides for the needs thereof."[4]

The portions allotted to us in life will of course differ. Some attain riches and some struggle for subsistence. But ultimately there is no objective standard for affluence. Affluence is only in one's art of being content with what one has. As the ethical tractate Abot declared it: "Who is rich? He who is content with his lot."[5]

ENVY, JEALOUSY AND PRIDE

The rabbis decried envy and jealousy, in which a person, out of discontent with his portion, begrudges the good fortune that has come to others. Envy and hatred of one's fellow-man were cited by the rabbis as vices that "take a man from the world."[6] One of the rabbis was accustomed to offer a daily prayer: "May it be acceptable before Thee O

Lord my God and God of my fathers, that no hatred against us may enter the heart of any man, that no hatred of any man enter our heart, that no envy of us enter the heart of any man, nor the envy of any man enter our heart ..."[7]

The rabbis were equally emphatic in denouncing pride. "Humility," one of the rabbis said, "is the greatest of all virtues."[8] A person who is puffed up with an arrogant spirit is as though "he had worshipped idols, denied the basic principles of religion, and committed every kind of immorality ..."[9] Arrogance is not only an evil trait because it hurts other people. It is equally injurious to its own possessor for it sends him on a road that will inevitably lead to frustration. The rabbis generalized thus: "Whoever runs after greatness, greatness will elude him; whoever flees from greatness, greatness will pursue him."[10]

THE MEANING OF GOOD WILL

The proper disposition of man toward his neighbor is an unreserved good will. The ethical tractate Abot reiterates this demand repeatedly. Matthew ben Heresh taught: "Be the first to offer cordial greetings to every man." Shammai was the author of a similar maxim: "Receive every person with a glad disposition." Ben Zoma was wont to say: "Who is deserving of honor? He who honors other people." Rabbi Eliezer urged: "Let the honor of your friend be as dear to thee as thine own." Rabbi Hanina ben Dosa declared: "He who pleases the spirit of man, will also please the spirit of God; and he who does not please the spirit of his fellowman, will not please the spirit of God either."[11]

The Talmud tells many anecdotes to illustrate the need of being ever vigilant to maintain one's good will toward others. One of these is the case of Rabbi Elazar ben Simeon who had become corrupted with pride because of his great learning and then came to look with disdain on other people. He once rode leisurely on his donkey at the edge of the river and felt especially pleased with himself, when he noticed a very ugly-looking person coming his way. The latter greeted him but he did not reply. Instead he asked whether all his townsmen were as ugly as he. The stranger's comeback was: "I don't know, but I suggest you go to my Maker and tell him: 'How ugly is this vessel you have made!'" At once the rabbi was aware that he had sinned. He descended from his donkey and bowed before the stranger and asked his forgiveness. The latter refused and he followed him with his entreaties to the entrance of the town. The people turned out in large numbers to welcome Rabbi Elazar and the stranger reported to them the incident. They joined in the entreaties, and the latter then agreed to accept the apology, on the understanding "that he shall never again act thus."[12]

Even if one have a genuine grievance toward his neighbor, he ought not to respond with hatred. The Talmud cited the case of a man cutting with one hand and inadvertently hurting the other hand. "Shall he in retaliation cut the hand that wielded the knife?" We are all part of one another and the hurts we inflict on others really strike at ourselves, since our lives are interdependent.[13] There are other ways of coping with grievances—to speak with candor and through honest voicing of our grievances to bring about a reconciliation. Indeed, a lasting friendship depends on the regular rebukes that one administers to the other. "A love without rebuke is no real love."[14] It takes much in self-control to act with such magnanimity toward those who have wronged us. But it is in such self-control that true character reveals itself. The true hero, teaches the Talmud, is "one who converts an enemy into a friend."[15]

One's good will should be extended without limits. Even the sinner is entitled to it. A Talmudic anecdote illustrates this. "There were some lawless men living in the neighborhood of Rabbi Meir and they used to vex him sorely. Once Rabbi Meir prayed for their death. His wife, Beruriah, thereupon exclaimed: 'What do you take as the sanction for your prayer? Is it because it is written, Let sinners cease out of the earth? (Ps. 104:35) But the verse may also be rendered to mean, Let sin cease out of the earth. Consider, moreover, the conclusion of the verse: And let the wicked be no more. When sins shall cease, the wicked will be no more. Rather should you pray that they repent and be no more wicked.' Rabbi Meir offered prayer on their behalf and they repented."[16]

The Talmud includes many anecdotes to illustrate the extent to which one ought to be patient with people. The hero in one such anecdote is Hillel. "Our masters have taught: A person should always be patient like Hillel and not quick-tempered like Shammai. Two men once made a wager that whoever would succeed in getting Hillel to lose his temper would win four hundred *zuz*. That day happened to be the eve of the Sabbath and Hillel was then washing his head. One of the men came to the door of the house and shouted, 'Is Hillel here? Is Hillel here?' Hillel wrapped himself, came out and asked him, 'What do you want, my son?' 'I have a question to put to you.' 'Ask it, my son.' 'Why are the Babylonians round-headed?' 'You have put an important question to me,' Hillel answered. 'The reason is that they have no skilled midwives.'

"The man left and after a short while returned, shouting, 'Is Hillel here? Is Hillel here?' The Rabbi wrapped himself, came out to him and asked, 'What do you want, my son?' 'I have a question to put to you.' 'Ask it, my son.' 'Why are the inhabitants of Palmyra bleary-eyed?' 'You asked an important question,' Hillel again replied. 'The reason is that they live in sandy districts.'

"The man went away, waited a brief while and again returned, shouting, 'Is Hillel here? Is Hillel here?' The Rabbi wrapped himself, came out to him and inquired, 'What is it, my son?' 'I have a question to put to you.' 'Ask it, my son.' 'Why are the Africans broad-footed?' 'You have asked an important question,' Hillel once more responded. 'The reason is that they live in marshy districts.'

"The man said, 'I have many more questions to ask, but I am afraid of provoking your anger.' Hillel folded the wrap about himself, sat down and said, 'Ask all that you desire.' 'Are you Hillel whom people call Prince in Israel?' 'I am.' 'If so, may there not be many like you in Israel.' 'Why, my son?' 'Because through you I have lost four hundred *zuz*.' The Rabbi then told him, 'Be careful, Hillel is worthy that you should lose through him four hundred *zuz* and still another four hundred *zuz*. But Hillel will not lose his temper.'"[17]

A good man is a peace-loving man. It was Hillel who extolled the virtue of peace in these words: "Be of the disciples of Aaron, a lover of peace and a pursuer of peace, one who loves mankind and draws them nearer to the Torah." According to Rabban Simeon ben Gamaliel, peace is one of the three pillars that sustain civilization, the other two being justice and truth.[18] Peace is the condition for the enjoyment of all other blessings. There may be food, there may be drink, but "if there is no peace there is nothing." Thus the rabbis advised people to shun quarreling. One who can exercise such restraint "will escape a hundred evils." The quarrelsome person who readily gives vent to his anger "will destroy his home."[19]

The strife between people arises often through misunderstandings. If we only knew all the circumstances under which our neighbor acted, we might understand and readily forgive that which caused our resentment. The rabbis accordingly recommend that we be cautious in judgment and that we accord each person the full benefit of our doubt. Hillel said: "Do not judge your neighbor unless you have been put in his place." Joshua ben Perahyah generalized: "Judge every man by the scale of merit."

As a helpful attitude to the maintenance of good relations with people, the rabbis suggested: "If you have done your neighbor a little wrong, let it be in your eyes great; if you have done him much good, let it be in your eyes little; if he has done you a little good, let it be in your eyes great; if he has done you a great wrong, let it be in your eyes little."[20]

The admiration of the rabbis for the peacemaker is clearly revealed in the following story: "A rabbi was standing in the marketplace when Elijah appeared to him. The rabbi asked him, 'Is there anybody in this marketplace who will have a share in the life of the world to come?' Elijah answered that there was not. Then two men appeared, and Elijah said, 'These two will have a share in the world to come.' The Rabbi asked them what they had done to earn such distinction. They answered, 'We are merrymakers; when we see people troubled in mind we cheer them, and when we see two men quarreling we make peace between them.'"[21]

THE IMITATION OF GOD

A person should actively pursue the welfare of his neighbor. The rabbis rooted this demand in man's duty to imitate God's providence. Thus the Talmud expounds: "What is the meaning of the verse, 'Ye shall walk after the Lord your God' (Deut. 13:4)? It is to follow the attributes of the Holy One blessed be He: As He clothed the naked (Gen. 3:21), so do you clothe the naked; as He visited the sick (Gen. 18:1), so do you visit the sick; as He comforted mourners (Gen. 25:11), so do you comfort those who mourn; as He buried the dead (Deut. 34:6), so do you bury the dead." The same thought is expressed in the Midrash: "As the All-present is called compassionate and gracious so be you also compassionate and gracious and offering thy gifts freely to all. As the Holy One, blessed be He, is called righteous (Ps. 145: 17) be you also righteous; and as He is called loving (ibid), be you also loving."[22]

THE MEANING OF BENEVOLENCE

The active concern for another person's welfare finds many expressions, but none is prized as much as *gemilut hasadim*, acts of loving-kindness or benevolence. Among the typical acts of loving-kindness mentioned in the Talmud are visiting the sick, hospitality to strangers, providing a proper outfit and dowry for a poor bride, caring for the orphaned. Highest of all is what we do for the departed such as attending a funeral and comforting the mourners.[23]

Talmudic literature abounds in the request to relieve the poor in their distress. But acts of benevolence are greater than almsgiving. The rabbis contrasted benevolence from almsgiving: "Greater is the benevolence than alms in three respects—almsgiving is performed with money and benevolence with personal service or money; almsgiving is restricted to the poor and benevolence applies to the poor as well as to the affluent; almsgiving applies only to the living and benevolence applies both to the living and the dead."[24]

The obligation to help the poor was an axiomatic element in Jewish morality. To the Romans it seemed strange, however. They treated the destitute with contempt, holding them in some ways responsible for their own distress. Occasionally Romans challenged the Jewish emphasis on the duty of helping the poor. The Talmud quotes one such discussion between Rabbi Akiba and Tineius Rufus, the Roman governor of Palestine: "Tineius Rufus asked, 'If your God loves the poor, why does He not provide for them? To cite a parable: Suppose a human king was angry with his slave, imprisoned him and ordered that he was not to be provided with food and drink; and then a person goes and feeds him and offers him to drink. When the king hears of it, will he not be angry with him?' Akiba replied, 'I will offer you a more appropriate parable: Suppose a human king was angry with his son, imprisoned him and ordered that he was not to be provided with food or drink; and then a person goes and feeds him and offers him to drink. When the king hears of it, will he not reward him?' We are called God's children, as it is said, 'You are the children of the Lord your God' (Deut. 14:1). Behold it was He who declared, 'Is it not to deal thy bread to the hungry and that thou bring the poor that are cast out to thy house?'" (Is. 58:7)[25]

Rabbi Akiba is the hero in another story which likewise extols our responsibility for the poor. "It was said of Rabbi Tarphon that he was exceedingly rich but did not give to the poor. Once Rabbi Akiba met him and asked, 'Would you like me to buy a town or two for you?' He agreed and offered him four thousand golden denarii. Akiba took them and distributed them to the poor. After a while, Rabbi Tarfon met him and asked, 'Where are the towns you bought for me?' Akiba took him by the hand and led him to the House of Study; he then brought a copy of the Psalms, placed it before the two of them, and they continued to read till they reached the verse, 'He hath dispersed, he bath given to the needy; his righteousness endureth forever' (Ps. 112: 9). Akiba exclaimed, 'This is the City I bought for you!' Tarphon arose, kissed him, and said, 'My master and guide, my master in wisdom, and my guide in right conduct.' He handed him an additional sum to distribute in charity."[26]

TRUTH IS THE SEAL OF GOD

Another great virtue extolled by the rabbis is truthfulness. "Truth," taught Rabbi Hanina, "is the seal of God Himself." Those who simulate in their speech were looked upon by the rabbis as idolators. Not merely fraud itself, but misleading a person in his opinions is condemned by the rabbis. The rule of the Talmud is: "It is forbidden to mislead a fellow-creature, including a non-Jew." "The Holy One, blessed be He," a Talmudic statement generalizes, "hates a person who says one thing with his mouth and is of another opinion in his heart." According to Rabban Simeon ben Gamaliel, truth is one of the three pillars on which the world rests; the other two are justice and peace.[27]

The rabbis condemned even the innocent lies which parents tell their children. These lies set an example in untruthfulness which children will in due time imitate. As one rabbi put it: "A person should not promise his child that he will give him something without giving it to him, for thus he teaches him to lie."[28]

The Talmud recounted with much admiration the exemplary honesty of some of its heroes. Rabbi Pinhas ben Yair and Rabbi Simeon ben Shetah figure in some of these stories. "It happened that Phineas ben Yair was living in one of the cities of the South, and some men who came there on business left two measures of barley in his possession and departed, forgetting all about the barley. He sowed the barley and each year stored the produce. After seven years had elapsed the same men returned to the town, and asked for their barley. He recognized them and asked them to take the entire produce." Another incident is related concerning Simeon ben Shetah. He had purchased a donkey from an Arab. His disciples noticed a gem hung from its neck, and they said, 'O, master, in you has been fulfilled, The blessing of the Lord maketh rich' (Prov. 10:22). He replied to them: 'I bought the donkey and not the gem.' He then proceeded to return it to its owner. The Arab, on getting it back, exclaimed, 'Blessed be the God of Simeon ben Shetah.'"[29]

THE PLEA FOR MODERATION

The man idealized by the rabbis is not the ascetic who shuns the world and its pleasures. It is rather the one who knows how to live within it in moderation. The world in all its fulness is a divine creation. Enjoying it is therefore a person's privilege, nay, his duty. The rabbis declared that a person is destined to give account to his Maker for all the good things his eyes beheld that he did not partake of. The rabbis commended the person who possessed "a beautiful home, a beautiful wife, fine furnishings." These put a person into "a happy frame of mind."[30]

The rabbis decried the ascetic's assumption of voluntary fasts as evil. According to the Babylonian teacher Samuel, he who indulges in fasting "is called a sinner." Another teacher, Resh Lakish, forbade fasting because it weakens one's body and thus lessens his services to God's kingdom. As a mark of disapproval, another teacher suggested giving the food shunned by the ascetics, to the dogs. The *nazirite* whose vow to reject wine is recognized as binding in the Bible (Nu. 6:1–4), the rabbis held to be a sinner, and they added: "If a person who withholds himself from wine is called a sinner, how much more so is one a sinner who withdraws from all of life's enjoyments."[31]

The rabbis were not unmindful of the dangers in indulgence to excess. Wine especially may be taken to excess and then it is injurious. Thus they warned: "Do not become intoxicated and you will not sin"; "when wine enters, sense leaves, when wine enters, the secret blurts out"; "one cup of wine is good for a woman, two are degrading, three make her act like a lewd woman and four cause her to lose all self-respect and shame."[32]

A rabbinic story portrays vividly the steps in degradation which a man walks when he gives himself to excessive drinking: "When Noah came to plant a vineyard (Gen. 9:20), Satan appeared before him and asked, 'What are you planting?' 'A vineyard,' Noah replied. 'What is its nature?' Satan continued. 'Its fruits are sweet whether fresh or dry, and wine is made of them, which gladdens the heart,' Noah answered. 'Come now, let us two form a partnership in this vineyard,' Satan proposed. 'Very well,' said Noah. What did Satan do? He brought a sheep and slew it under the vine; then he brought in turn a lion, a pig and a monkey, slew each of them and let their blood drip into the vineyard and drench the soil. Thus he hinted that before a person drinks wine he is simple like a sheep and quiet like a lamb before his shearers. When he has drunk in moderation, he is strong like a lion and feels as though there is none to equal him in the world. When he has drunk more than enough, he becomes like a pig, wallowing in filth. When he is intoxicated he becomes like a monkey, dancing about, uttering obscenities before all, and unaware of what he is doing."[33]

The study of Torah was regarded by the rabbis as the supreme good of life, and yet they cautioned that even our preoccupation with Torah must not displace our concern with our worldly obligations. "Torah is good," said the rabbis, "when combined with a worldly occupation."

The Talmud tells of Rabbi Simeon ben Yahai who had hidden in a cave for twelve years in order to elude the Romans who sought to arrest him. When he finally emerged from his hiding place, he noticed that people about him were going on with their usual affairs, plowing and sowing, and exclaimed: "They forsake the life of eternity and busy themselves with the life that is transitory!" A heavenly voice finally rebuked him: "Have you left your cave to destroy my world? Go back to it!"[34]

CLEANLINESS AND HEALTH

The Talmud urged the proper care of the body as an obligation which one owes toward himself. Cleanliness they held a basic prerequisite to good health. "Rinse the cup before and after drinking," recommended the rabbis. Similarly they cautioned, "A person should not drink from a cup and hand it to another, for it is dangerous to health." The Talmudists lived among people who were especially troubled with eye disease, still a common affliction in oriental countries. But the Talmudists blamed it principally on the lack of sanitary habits among the people. "Better a drop of cold water in the morning, and the washing of hands and feet in the evening than all the eye salves in the world."[35]

The rabbis looked upon the maintenance of bodily health as a religious obligation. This is made clear in the following anecdote, in which Hillel is once more the hero. When Hillel had finished a session of study with his pupils, "he accompanied them part of the way. They said to him, 'Master, where are you going?' 'To perform a religious duty,' he replied. 'Which religious duty?' they asked. 'To bathe in the bath-house.' 'Is that a religious duty?' they wondered. He answered them: 'One who is designated to scrape and clean the statues of the king which are set up in theatres and circuses is paid for the work and he associates with nobility. Surely must I who am created in the divine image and likeness, take care of my body!"[36]

The Talmud abounds in rules of health, some of which will continue to interest the modern reader. The rabbis cautioned against overeating: "Restrain yourself from the meal you especially enjoy, and do not delay answering nature's call." They urged sufficient sleep, which will do its best however only at night; late morning sleep was regarded as injurious. Above all they urged general moderation in living: "In eight things excess is harmful and moderation beneficial: travel, sexual intercourse, wealth, work, wine, sleep, hot water (for drinking and bathing) and blood-letting." It is interesting that the rabbis recognized that bodily illness often derives from psychic causes. Thus they listed fear and sin among the things which "weaken a man's strength." In the event of illness the rabbis urged that a physician be consulted, and they forbade people making their homes in communities that were without the services of a competent physician: "It is forbidden to live in a city that is without a physician."[37]

THE JURISPRUDENCE OF THE TALMUD

Talmudic law differs from other systems of jurisprudence in its all-inclusive character. It is not confined to the realm of social relations. It seeks to implement the entire range of values which are taught in Judaism, whether they derive from religion or morality. We may define its goal as the enforcement of those elements of doctrine and conduct that the rabbis deemed indispensable to the life of the individual or the community.

Talmudic law concerns itself with doctrine, but it does not establish dogmas that must be believed in as true. On the level of opinion great freedom existed in the Jewish community and individuals were allowed to follow their own inclination of heart and mind. There was ample literature expounding the basic conviction of Judaism and the very diversities of thought and interpretation were deemed a source of strength in Jewish tradition. Truth cannot be contained in one easy formula. Like the fire which breaks into many sparks, so does truth break into many fragmentary truths, which are caught by diverse human minds. Talmudic law centers on the discipline of action, but the actions which it prescribed were also a vehicle of doctrines that the rabbis deemed indispensable in their way of life.[1]

LAW, THEOLOGY AND RITUAL

Talmudic law recognizes two general categories of value. One is the duties which derive from man's relationship to God; the other is duties which derive from man's relationship to his neighbor. The laws dealing with man's relationship to God are, in a sense, the implementation of Jewish teachings in theology. They are intended to deepen man's consciousness of those doctrines through repeated actions in which they are enshrined. Thus Talmudic law ordains the recitation of the *shema* (Deuteronomy 6:4) affirming the unity of God, twice daily, morning and evening. It establishes a ritual of daily public and private prayer. It formulates the specific texts of the benedictions on partaking of various foods. Through these rituals man was to be made more keenly aware that he is living in God's world and that he must be ever grateful for the privilege of enjoying its manifold blessings. To accept what the world offers us without a thought of what we owe to God for it, marks a man an ingrate. As the rabbis put it: "It is forbidden a man to enjoy the things of this world without a prayer."[2]

The law governing man's relation to God often serves also as a precautionary measure to prevent the transgression of more fundamental principles or doctrines. The rabbis pictured the basic elements of religion and morality which they wanted their people to maintain as a kind of vineyard that must be fenced in against violators. This was one of the guiding rules of the men of the Great Assembly: "Build a fence to the Torah."[3]

The law as a "fence to the Torah" is clearly illustrated in the widely ramified rules bearing on idolatry. The cult of idol worship was widespread throughout the Roman empire, and its visible symbols, images of all kinds, dotted prominent sites in city and country. Surrounded by these manifestations of paganism on all sides, the Jews were in danger of contamination. The danger was met by Talmudic law which forged a mighty fence to protect the religious purity of Jewish life. It declared all idolatry, its symbols, the site where they were located and all activities associated with it, out of bounds for a Jew. Even the broken wood or metal that had ever been part of an idol was forbidden. A grove where an idol was situated was not to be entered, even for the innocent purpose of being shaded from the sun. The wine employed in idolatrous offerings was not to be used. Even a drop of it falling into another liquid would render it unfit for normal consumption.[4]

The rabbinic struggle against idolatry was not a novel phenomenon in Jewish tradition. It appears in the Bible where it was directed against earlier forms of this religious primitivism. It is a continuation of one of the permanent characteristics of Judaism, its battle against the artistic glorification of the blasphemous error which reduced God to finite form. The discouragement of painting and sculpture in classic Judaism derives from this struggle against error made more palatable through beautiful representation. The rabbis fought an important episode in this struggle, and they achieved their victory through law.

The law which governs man's relation to God possessed qualities of adaptability, as did law of human relations. And it responded to the pressures of the circumstances under which it was to be lived. This is well illustrated in the law which forbids travel on the Sabbath.

The Sabbath was instituted in Judaism for a dual purpose. It was to be a memorial to creation, to recall to us the divine source of all existence. It was likewise endowed with social significance, to rest the bodies and minds of men, a goal that was inspired by the remembrance of the emancipation from Egyptian bondage. The measures by which the Sabbath was to be commemorated were many, and among them was the rule against travel. An examination of this rule in all its wide ramifications reveals the profound religious and moral ends which the rabbis sought to accomplish by it, and the fine line of development through which its basic elements finally emerged.

The prohibition to travel on the Sabbath is derived from the verse in Ex. 16:29: "Abide ye every man in his place; let no man go out of his place in the seventh day." Originally directed at the gatherers of manna in the wilderness, this verse was seen in a more general light, as an interdiction of all movement on the Sabbath beyond one's domicile.

Rabbinic sources offer us two general reasons for the objection to travel, both related to the goal of liberating man on the Sabbath day from labor as well as anxiety and distraction. The first consideration is expressed in the principle of *tehumin*, the need of fixing one's domicile in a particular place, and then limiting one's motions within a prescribed radius of that place. The Sabbath experience depended on keeping the family together within the atmosphere of the home, and the home had to be fixed in space, even as the Sabbath was fixed in time. On that day, man was therefore to confine his life to the home and its surroundings.

The original interpretation of the Biblical verse was literal, and the place of permissible movement was confined to the home plus an additional 2000 cubits. The tendency to socialize the Sabbath finally wrought a change in interpretation and the home was then taken in the widest possible sense, to include one's city, supplemented by the usual radius of 2000 cubits of additional movement. The terminus of allowed movement by an additional provision of the law, could, moreover, be pushed farther away when necessary through an *erub*, a conscious designation of the desired place outside the city as part of one's home, by depositing there some food as a token of home. A traveller who chanced to be away from a city at the advent of the Sabbath could, by an act of conscious designation known as *kinyan shebitah*, fix his home anywhere and then he was free to move within the 2000 cubit radius of that place.

Travel on the Sabbath by riding an animal was also forbidden for the additional reason of seeking to avoid involvement in incidental labor, such as possibly cutting down a twig in order to prod the animal on its way. There is also the suggestion that one who rides an animal might easily move beyond the confines of the *tehum* and cross the area around the home which is the zone of allowed movement on the Sabbath.

There are other elements in the law of the Sabbath which regulate movement, but they all testify to the same underlying goal. The rabbis did not seek arbitrarily to stifle the free movement of life. They sought to reject tension, undue exertion. They sought to mold the Sabbath into a day of serene, relaxed living. Thus they banned the *pesia gasa*, the hurried walk of the busy days of the week. The Tosefta generalized: "One may not run on the Sabbath to the point of exhaustion, but one may stroll leisurely throughout the day without hesitation." The Sabbath was to be a day of peace, and the halakah was engaged in fashioning the usual rabbinic fence that was to keep man from crossing over into the domain where the world and its cares stood ready to devour his serenity and his rest.

The Sabbath law was as flexible as every other branch of the halakah. Under some circumstances the prohibition against riding was waived, simply because other values at stake were deemed more pressing. Thus it eventually ceased to operate altogether in the case of ocean travel. The difficulty of pacing travel in such a way as to avoid being on the boat on the Sabbath was clearly the most significant factor. It would have paralyzed movements from Palestine to other parts of the world, which in many cases depended on schedules beyond the control of the individual passengers. In some instances the journey as a whole was of more than a week's duration, and it was clearly impossible to halt the ship for the Sabbath observing passenger.

That the rabbis originally looked upon ocean travel as included in the category of prohibited movement is manifestly clear from our sources. Thus the Talmud provides: "One must not undertake a boat voyage less than three days prior to the Sabbath. ... On the other hand, the short distance from Tyre to Sidon one may undertake even the day preceding the Sabbath."

91

In time the law reckoned with life and the formula was eventually worked out, allowing even the boarding of the ship on the Sabbath itself, provided one had deposited there some of his belongings, thereby designating it as his home for the Sabbath through an act of *kinyan shebitah.*

Travel on land, too, was in some exceptional cases suspended in consonance with other considerations, deemed even more pressing than Sabbath rest. Thus a witness testifying as to the appearance of the new moon—a vital consideration in the then current system of calculating the calendar—was permitted to travel on the Sabbath. He was to come riding on an animal even on the Sabbath day.[5]

LAW AND A JUST SOCIETY

The underlying goals of the law which derives from the relationships between man and man are more apparent. They seek to create a just social order that shall liberate man from arbitrary impediments to his growth. But the law of the Talmud does not consider itself as an impartial umpire that is to keep individuals within their respective spheres, without encroaching upon one another. "One who asserts what is mine is mine, and what is yours is yours, is only of medium ethical stature," according to the Talmud. There is even an opinion that such a standard corresponds to the ethics of the wicked city of Sodom.[6] The standard commended by the rabbis is the willingness to bend self-interest in acts of helpfulness toward others. And Talmudic law reflects this higher standard. It does not seek to balance self-interests. It seeks to bend the enterprises of society toward acts of welfare for the common man, especially for the underprivileged members of the community.

The standard of welfare which the Talmud recognized as ideal is total self-identification with the needs and aspirations of one's fellow-man. The Talmud calls it the standard of saintliness. The Mishnah defines it thus: "What is mine is thine and what is thine is thine is a *hasid,* a saintly man."[7] The standard of saintliness was not a practical standard by which men could order their lives in society. It projects an ideal which most men could not attain. The law crystallized at a moral level below this, but the ideal of saintliness played a tremendously vital role in rabbinic law. It proclaimed that the law in itself does not exhaust the moral ideal. It enabled men to judge their conduct by an ideal which, precisely because it was unattainable, could ever serve as a source of vital self-criticism and as a spur to new moral endeavor.

The recognition that the law did not realize the highest moral ideal led to a demand that men go beyond the limits of the law in their dealings with each other. This is clearly conveyed in the rabbinic interpretation of the verse in Exodus 18:30, "And thou shalt make them know the path they are to walk in and the work they are to do." "The path they are to walk in" according to Rabbi Elazar of Modein, refers to the law, while "the work they are to do," he continues, refers to acts of saintliness "beyond the measure of the law."[8] The rabbis cite various cases in which people of moral sensitivity acted on a higher standard than the one called for by the law, and their conduct is hailed as exemplary.[9]

Those actions "beyond the line of the law," as the Talmud calls it, constituted a free zone in which individuals expressed their generosity and love for their fellow-men, without compulsion from outside sources. The Talmud hailed this free zone of moral action as the very foundation of a good society. A community in which men are content to hew to the strict letter of the law was devoid of the moral cement that gives a social order stability and enables it to survive. "Jerusalem was destroyed," according to Rabbi Jananan, "because her people hewed strictly to the letter of the Torah."[10] It is actions beyond the law that give evidence of a vibrant morality and save the law itself from becoming a soulless formalism devoid of feeling and vitality.

MORAL PRESSURES ON THE LAW

The standard of saintliness was important not only for the individual in keeping alive for him the underlying moral impulses which the law in itself could not fulfill. It acted as a pressure on the law, forcing it to move forward to new frontiers of human service. The Talmud gives evidence of a continuously growing program of welfare legislation, in which ever wider sectors of social life were brought under the control of a law, whose motivating impulse was the welfare of the common man. Thus the law empowered the community to assume responsibility for elementary education and poor relief. It authorized the supervision of weights and measures, and of fair wages and prices to prevent unethical business practices.[11] The law compelled children to provide for the maintenance of parents, even as parents were compelled to provide for the maintenance of children.[12]

The law forced a person to help his neighbor where it was clear that he himself would not lose by it. Thus, heirs dividing land that had come to them by inheritance were expected to consider that one among them owned land contiguous to the parcel to be divided and to give him his share near his own land. The Talmud generalized: "We coerce against the standard of Sodom." A person did not have the absolute right to be mean.[13]

The pressure of a higher moral standard inspired the Talmudic liberalization of the Jewish criminal code. Capital punishment is provided in the Bible for a variety of crimes. But the rabbis, as we have already noted, found capital punishment reprehensible, and they rendered it almost inoperative by hedging it with conditions that made of the old law a dead letter. Thus they insisted that the commission of the culpable act must be preceded by a warning and by an expression of defiance on the part of the criminal in the face of that warning.[14] And the Mishnah declares explicitly, "A Sanhedrin which decides a verdict of death once in seven years is called murderous. Rabbi Elazar ben Azariah said, even if only once in seven years. Rabbi Tarphon and Rabbi Akiba said: 'If we were members of the Sanhedrin, there would never be a verdict of death.'"[15]

The growth of Talmudic law, in all its aspects, was for the most part, we have already noted, the work of judicial interpretation rather than of formal legislation. The rabbis who were called upon to administer the old law reckoned with the conditions under which it was to be applied. And if they thought the mechanical application of precedent in conflict with the demands of equity, they resorted to reinterpretations which withdrew the new case from the old category into which it seemed, by the rules of formal logic, to fall. The case so decided then became precedent for parallel situations.

The judge served in effect as a creator of law and not only as its interpreter—a phenomenon which has been duplicated in every system of jurisprudence. Thus the limitation of capital punishment to instances which satisfied the qualifying circumstances was an act of judicial interpretation. But it set a precedent which broke new ground in the entire range of Jewish criminal law.

THE BASIS OF LEGAL CONTROVERSY

It goes without saying that these far-reaching judicial interpretations did not proceed with universal concurrence. Considerations of equity are ultimately subjective in character and they will reflect the diverse hearts and minds in which they occur. This is the principal reason for the marked presence of controversy in the Talmud. The rabbis were not contentious for contention's sake. They disagreed as do the judges on any judicial tribunal. They were simply offering diverse reactions to the problems of life, born of diverse backgrounds and of those intangible diversities of temperament, character and outlook, which naturally divide men from one another. Thus, the decision against capital punishment was challenged by Rabbi Simeon ben Gamaliel who defended the old law as an indispensable deterrent to crime. The reform proposed, he argued, would "cause an increase of bloodshed in Israel."

The differences of opinion among the Talmudists are rot always indications of genuine disagreement. They are rather, in many cases, the varying customs and usages which derive from their respective backgrounds. Thus Rabbi Eliezer, who was an aristocrat, exempted arms from the prohibition of carrying unnecessary objects on the Sabbath. He regarded them as ornaments and they were to be worn as a normal part of a person's apparel. His colleagues, representing the point of view of the common people, forbade it. Citing the prophetic contempt for war and its implements, they branded the wearing of arms as a "disgrace".[16]

A similar difference, deriving from the diverse backgrounds of the rabbis, is offered us in the definition of the time when the Shema is to be recited, evening and morning. The Bible defined the time as "when thou liest down" and "when thou risest up." Rabbi Eliezer ben Hyrcanus, reflecting his rural background where it is customary for people to retire early and rise early, sets the time in the evening from sunset to the end of the first watch of the night, or nine o'clock. In the morning he sets the time from the appearance of the first streaks of light till sunrise. His colleagues, reflecting an urban practice, permit the Shema in the evening until midnight and in the morning until nine o'clock.[17]

TALMUDIC LAW AND THE STATE

The rabbis who created Talmudic law were the religious representatives of the Jewish community; they were not functionaries of the state. Prior to the destruction of the Temple in 70 C.E. the state was intermittently under the influence of the Pharisees, the forerunners of the rabbis who were the great builders of Talmudic law. The most influential molders of policy, however, were Sadducees. Pharisaic interpretation had a great moral force among the people, and to that extent exerted pressure with which the state had to reckon. We have a record of Alexander Jannai, king and high priest, proceeding to perform the succot ritual in the Temple according to Sadducean ritual, whereupon the assembled worshippers demonstrated in protest.

Talmudic law came into its own after the destruction of the Temple. In the limited autonomy enjoyed by the Jewish community in Palestine and in Babylonia, Jewish law was given far-reaching scope; and that law was the law as interpreted and administered by the rabbis. Yet in many cases the state asserted its own sovereignty to supersede the internal law of the Jewish community. The rabbis advised conformity. The Babylonian teacher Samuel ruled explicitly: "The law of the state is law."[18] This became the basic rule governing the Jewish attitude toward his obligations as a citizen. His own law retreated to make room for the law decreed by the state of which he deemed himself a part.

The Talmud drew a line, however, as to how far the accommodation of Jewish law to the state was to proceed. Where the state sought to violate basic principles of morality and faith, its law was to be resisted. As the Midrash declared, commenting on the verse: "I counsel thee, keep the king's command and that in regard of the oath of God" (Eccles. 8:2): "The Holy One, blessed be He, said to Israel, 'I adjure you that if the government decrees harsh decrees, rebel not against it in any matter which it imposes upon you, but keep the king's command; if, however, it decrees that you annul the Torah and the precepts, do not obey.'"[19]

The dilemma here posed became a real issue during the reign of the Emperor Hadrian. As part of the Roman empire, Palestine and her Jewish community became subject to imperial law. The edict of Rome proscribed all the practices of Judaism on pain of death. The rabbis met the challenge by calling for conformity, with the exception of the three fundamentals, the laws against idolatry, immorality and murder. A person was to suffer martyrdom rather than violate these in conformity to the unjust will of the state. As the rabbis put it: "Nothing must stand in the way of self-preservation, except idolatry, immorality and bloodshed." Rabbi Ishmael limited the demand for martyrdom in the case of idolatry, to its public profession. In privacy he called for compromise even in this instance, rather than suffering martyrdom.[20]

LAW AND INWARDNESS

Law is a discipline which governs action. But the rabbis were keenly aware that the inner man is more important than the deed through which he expresses himself. "The Holy One, blessed be He, is concerned above all with what is in man's heart."[21] For a person may conform to the demands of the law, and remain inwardly corrupt. And similarly a person may in the midst of a life of wrongdoing go through an intense experience of inner change that leaves him a noble character. "One man earns his place in the world," Rabbi Judah the Prince, once reflected, "through the efforts of many years, and another earns it in one hour."[22] Indeed, Rabbi Abahu ranked the penitent even above the man who had never sinned.[23]

The decisive hour of repentance may transform a sinner into a saint. But the rabbis distinguished as to its sufficiency between the relations of man to God and the relations of man to man. Repentance will wholly clear a person for transgressing laws expressive of our relations to God. More is, however, required in the case of transgressions of the law of human relations. The aggrieved person must be appeased. Thus the Mishnah declares: "Transgressions between man and God may be atoned on the Day of Atonement, but transgressions between man and man will not be atoned on the Day of Atonement until one has appeased his fellow-man."[24]

It is significant, however, that the rabbis limited the scope of this required appeasement, in order not to place a discouraging burden on the would-be penitent. Thus one who had robbed a beam and built it into his house, was not required to damage his building by tearing out the beam to return it. It was deemed sufficient if he returned the value of it.[25]

The recognition of inwardness as a factor in law led to far-reaching consequences in the jurisprudence of the Talmud. It led to the demand that in the application of law we reckon not only with the letter of the law, but also with the manifest intention of those responsible for its enactment. This is well illustrated in the Talmudic interpretation of the Sabbath law. Thus, according to the Bible, violators of the Sabbath law by performing forbidden labor, whether in error or ignorance, were required to bring a sin-offering as a sacrifice. But what if a person committed, in one span of forgetfulness, a number of Sabbath violations, either on the same Sabbath or spread over a number of Sabbaths? How many sin-offerings was he to bring? The Talmudists ruled that the sin-offering was obviously intended to atone for negligence, and not for the labor as such. Since only one span of forgetfulness was involved, only one sin-offering was to be brought.

The Talmudists demanded also that the law reckon with the intention behind the deed, and not merely with the deed itself. Thus they absolved a person from all guilt if a stone thrown by him accidentally fell upon some one and injured him. They also absolved a person from the charge of murder if, intending to kill an animal, he missed his target and killed a human being. Where a person intended to kill a human being and missed his target, killing instead another human being, there was a difference of opinion among the Talmudists. Rabbi Eliezer regarded the act as murder; Rabbi Simeon did not.

The Talmudists allowed certain fulfillments of the law to the free play of spontaneous decision. No fixed measure was given for the area on the corner of each field which was to be left as a beneficence to the poor. Nor was there a fixed measure for the offering of the first fruits of the harvest that was to be a gift for the priest, or for the offerings brought on appearing at the Temple during the pilgrimages on the three major festivals, or for the practice of charity and the study of Torah.

The most significant expression of spontaneity in Talmudic law was the recognition of a wide range of authority for local custom, or minhag, as it was called. Local communities, trades, and even family groups often adopted measures to govern their religious or social life, or commercial transactions. These arose spontaneously, in areas which were not covered by the law. The rabbis invested these customs or *minhagim* with authority, and demanded compliance with them. Indeed, where a law clashed with a deeply rooted custom, they often gave precedence to custom.[26]

THE LAW IN MESSIANIC TIMES

The rabbis envisioned an even wider scope for religious and moral inwardness to be attained as history reaches its final unfolding. They anticipated that inwardness would eventually vanquish law altogether. In Messianic times when men will have learned the true lessons of the love of God and the love of man and feel that love deep within themselves, the law will no longer be necessary. The cult of worship by which we now express our relation to God and the apparatus of justice by which we now administer the law of human relations, will then become obsolete. For it will then be possible to depend on human spontaneity, expressing ennobled human characters, to suggest the right action in every situation without the discipline of law to channel it. "The laws," the Talmud declared, "will become obsolete in the hereafter."[27]

In the present stage of human immaturity, however, the law is an indispensable guide to action. It is, moreover, a preparation for the next stage of civilization, when the law which has come "to ennoble the lives of men"[28] will have done its work. A new human race will then arise to live on the level of true inwardness, in free gestures of adoration of God and in an all-embracing love for their fellow-men. The rabbis expressed this vision in their conception of the three stages of human history. The first is the stage of "chaos", before the leaven of a divine law has begun to work in the world; the second is the stage of "Torah"; and the last is the stage of Messianic liberation and enlightenment which will finally bring man to his pre-ordained destiny.[29]

HUMAN WISDOM IN THE TALMUD

The world outlook of the rabbis is often an elaboration of some revered utterance by a Biblical writer or some other master of tradition; occasionally it is the fruit of some new inspiration that has carried its recipient into the ranks of the creative builders of Jewish thought. There is, however, an additional force that is represented in their pronouncements—it is the common human wisdom, which men have always distilled out of the general experiences of life.

DREAMS AND THE SUBCONSCIOUS

The rabbis were shrewd observers of human nature in action. They were aware of the subtle life of the mind, recognizing that conscious experience is only a phase of a larger world in which we have our being. The rabbis were of course far away from the insights of modern psychology. Yet they recognized fully that the subconscious performs its delicate operations—as in dreams for instance—out of the materials furnished by the conscious, out of the hopes and fears that agitate the mind in normal life.

The Talmud cites a variety of notions concerning the significance of dreams. Among them is the recognition that dreams are nothing but elaborations of thoughts dwelt upon in hours of consciousness. Thus R. Samuel ben Nahman on behalf of R. Jonathan said: "Dreams are representations of thoughts on which one continues to meditate in one's wakefulness." This conception of dreams is forcefully presented in a reported conversation between the Roman emperor and Rabbi Joshua ben Hananiah: "'You claim to be wise men,' the emperor said to the rabbi. 'Tell me then what I shall see in my dream.' He replied, 'You will see the Persians (Parthians) enslaving you, despoiling you and making you pasture unclean animals with a golden staff.' The emperor continued to reflect on this all day and at night dreamed of it." The same Talmudic text records a similar experience on the part of the Parthian king, Shapur, with Samuel as the rabbi suggesting the subject of the dream.

The rabbis recognized that dreams are often pure fancy. Yet they felt that even in the seemingly incomprehensible dreams there are vital references to conscious experience. They sought a key to unravel the veiled allusions of our dreams which employ a language of symbols that need interpretation.

The interpretation of dreams was popular among the Talmudists. But they suggested that often it is the interpretation which becomes suggestive to the conscious mind of hopes or fears, which then condition the direction of our lives. The rabbis therefore cautioned people not to become unduly disturbed by dreams: "Dreams have no importance for good or ill."[1]

THE HEART IS SOVEREIGN

The rabbis were impressed with the profoundly important role that emotions play in life. The heart, which they looked upon as the seat of emotion, was regarded by them the principal source of control over all human actions. "All of man's bodily organs are dependent on the heart," was a Talmudic dictum. It is the heart therefore which may be said to carry responsibility for whatever we do in life. Thus one rabbinic comment offers us the sweeping generalization: "The heart sees, hears, speaks, walks, falls, stands, rejoices, hardens, softens, grieves, fears, is broken, is haughty ... persuades, errs, fears, loves, hates, envies, searches, reflects. ..."

The rabbis prized highly the ability of some people to control their emotions. To control one's emotions and to bring life under the directing voice of reason was regarded by the rabbis as the mark of true heroism. "Who is a hero?" one rabbi asked in the ethical treatise Abot. His reply was: "He who controls his passion."

HABIT AND CHARACTER

The Talmud abounds with statements which clearly recognize the dominant role of habit in human conduct. Character is to a large extent a pattern of behavior formed by habit. Our conduct is always conditioned by the chain of preceding actions, which predispose us to one way of life or another. "A good deed," according to the ethical treatise Abot, "leads to another good deed, and the consequence of one transgression is another transgression."

Habit is a mighty fortification of the good life. For once we habituate ourselves to noble living, the normal bent of our character will incline us toward the right deed in the particular situation confronting us. And any attempt to deviate from what has become the norm for our life, will be met with inner resistance. But the rabbis warned that a pattern of behavior once formed, is not necessarily of permanent duration, and that the sensitivity to these deviations from the norm will gradually wane, as the act is repeated. As the Talmud puts it: "When one transgresses a commandment and repeats the offense he feels no further restraint."

The rabbis consequently urged caution in behavior, warning people against even seemingly trivial slips in conduct. These slips are grave, for they predispose man to a course from which he may find it difficult to turn back. "He who violates a seemingly trivial statute will eventually violate a weighty one." The only sound advice is thus constant vigilance: "Avoid even a minor transgression lest it lead you to a major one."[2]

A MAN WEARS MANY MASKS

The Talmudists recognized that human character is often hidden beneath appearances, and that men may simulate virtues they do not really possess. But they suggested situations which will reveal what is intrinsic in man. Pretense, they explained, will disappear in situations involving money matters, in moments of anger or by the way a man takes his liquor. As R. Ilai tersely phrased it: "You can recognize a person's real character by his wine cup (koso), his purse (kiso), and his anger (kaaso)."

The discussions of the rabbis reveal the recognition of the immense power which the craving for material possessions exercises over people: "No man departs from this world with half his cravings satisfied. When he has attained a hundred, he desires two hundred."

The rabbis commented sadly on the tendency of people to cultivate well-to-do friends, and then to desert them when they suffer a reversal in fortune. "At the gate of the enterprising shop, there are many friends and brothers. At the gate of a shop in decline there are neither brothers, nor friends." Raba was even more pointed in his observation: "When the ox is fallen the knife is sharpened."[3]

The tendency of people to hide beneath a mask of pretense and falsification creates an element of uncertainty in every human relationship. It leads to deceit, and to the incompatible claims of litigants. The rabbis therefore sought a clue to the workings of the human mind which would enable us to probe through the false claim and to discover the true facts in a given situation. The Talmud records a number of principles which guided them in their deliberations.

It was taken for granted that a squatter's occupancy of any property would normally be challenged by its rightful owner within a three-year period of time. And if no such challenge developed in that time, the occupant may be presumed to be there by right, even though he might not have any documentary evidence to establish his rights. A liar was always presumed to fabricate the lie that would be to his greatest advantage. Greater credence was therefore to be placed to a plea yielding a lesser advantage than what was possible under the circumstances. It was assumed that falsification was less likely when the claimants confronted each other. Another important presumption was that a person does not normally pay his debts until they fall due. A person was assumed to be blind to his own shortcomings.

The application of these "presumptions" concerning human nature was at times challenged by the rabbis. For these are not iron-clad rules inexorably at work in all instances. Many a man may deviate from common procedure. This is clearly indicated in the following discussion: "Resh Lakish laid down the ruling: If a lender stipulates a date for the repayment of a loan, and the borrower pleads (when the date of payment arrives) that he paid the debt before it fell due, his word is not believed. It is enough if a person pay when his debts fall due. Abaye and Raba both concur in saying that it is not unusual for a man to pay a debt before it falls due; sometimes he happens to have money, and he says to himself, 'I will go and pay him, so that he may not trouble me.'"

The rabbis were fully aware of individual differences among people, and they often sought some indication of the mind of the particular parties involved in a litigation. This is well illustrated in the following case: "A certain Ronya had a field which was enclosed on all four sides by the fields of Rabina. The latter fenced them and said to him: 'Pay me toward what I have spent for fencing.' He (Ronya) refused. Then he asked, 'Pay toward the cost of a cheap fence of sticks.' But Ronya again refused. He continued, 'Then pay me toward the cost of a watchman.' Ronya still refused. Then one day Rabina saw Ronya gathering dates, and he said to his manager, 'Go and snatch a cluster of dates from him.' He went to take them, but Ronya shouted at him. Whereupon Rabina said, 'You show by this (shouting) that you are pleased with the fence. If it is only goats (you are afraid of), does not your field need guarding?' He replied, 'A goat can be driven off with a shout.' But he said, 'Don't you require a man to shout at it?' He appealed to Raba who said to him, 'Go and accept his last offer …'" (to pay toward the cost of a watchman).[4]

ON THE PSYCHOLOGY OF WOMEN

The Talmud quotes many proverbs that deal with the power of the sexual attractions of men and women. "No one is immune to the ravages of an illicit attraction." "There is only one real cause of jealousy among women—sex appeal." The Talmud recognized a woman's love for finery and personal adornment. "A woman is concerned principally with her appearance," one Talmudist observed. "And the greatest pleasure a man can give his wife is to clothe her in fine garments."

The love of self-adornment among women is more elaborately treated in the following passage: "These are the treatments of women—treating the eyes with kohl, curling the hair into ringlets, and rouging the face. The wife of R. Hisda used to adorn the face of her daughter-in-law. R. Huna ben Hinena once sat in the presence of Rab Hisda and, observing his wife apply the beauty treatment on her daughter-in-law, said, 'It is only permitted in the case of a young woman, not an old one.' He replied, 'By God, it is even permitted in the case of your mother and grandmother, and even if she stood on the brink of the grave; for as the proverb put it, "At sixty or at six, a woman runs after the sound of the timbrel."'"[5]

The rabbis record other observations on the psychology of women: "God endowed a woman with keener judgment than man"; "women are compassionate"; women are "querulous and garrulous"; women have an affinity for the occult and they go in "for witchcraft."[6]

The rabbis recognized the subtle influences of a woman in directing the life of her husband. This is told dramatically in the Midrash: "A pious man had been married to a pious woman but, being childless, they were divorced. He then went and married a wicked woman, and she made him wicked. The divorced woman proceeded and married a wicked man and she made a good man out of him. It thus follows that everything depends upon the woman."[7]

The same Midrash tells another tale which extols modesty as a woman's noblest virtue, at the same time alluding to common weaknesses in a woman's character. The text on which this homily is based is Gen. 2:21, where it is told that Eve was formed from one of Adam's ribs: "God deliberated from which part of man to create woman. He said, 'I must not create her from the head that she should not carry herself haughtily; nor from the eye that she should not be too inquisitive; nor from the ear, that she should not be an eavesdropper; nor from the mouth that she should not be too talkative; nor from the heart that she should not be too jealous; nor from the hand that she should not be too acquisitive; nor from the foot that she should not be a gadabout; but from a hidden part of the body that she should be modest."[8]

EDUCATION AND HUMAN NATURE

The psychological notions of the Talmudists had their most fruitful application in the field of education. The rabbis recognized individual differences among students, and they demanded that the educational process reckon with those differences. Some of these differences are discussed in the ethical treatise *Abot:* "There are four types among students. One comprehends readily but forgets readily—his advantage is nullified by his disadvantage; one is slow to comprehend but also slow to forget—his disadvantage is nullified by his advantage; one comprehends readily and forgets slowly—his is a good portion; one is slow to comprehend and quick to forget—this is a bad portion."

Another classification, also cited in the treatise Abot, deals with the relative reactions of students to knowledge given them: "There are four types among those who sit before the wise: the sponge, the funnel, the strainer, and the sieve. Some are like the sponge which absorbs everything; some are like the funnel which takes in at one end and lets out at the other; some are like the strainer which allows the wine to go out and retains the dregs; some are like the sieve which lets out the bran and retains the fine flour."

A more fundamental differentiation of students, on the basis of aptitude, is given in the Midrash: "Said R. Judan ben Samuel, 'The Torah, given by the Eternal, was offered us only in relative measure … Some quality for the study of Bible; some for the Mishnah; some for the Talmud; some for Aggadah; and some for all of these.'"[9]

A variety of other material in educational psychology is scattered in the writings of the Talmud. The importance of motivation and interest in education is recognized in the comment of Rabbi Judah the Prince: "A person can learn only those portions of the Torah which his heart desires." A combination of teacher's aloofness with a friendly interest in his students is demanded in the aphorism: "Always push the student away with the left hand and draw him near with the right." Teachers were urged to lay great stress on repetition. Rabbi Elazar was said to repeat his lesson four times. Students were urged to study out loud and place themselves in a position where they could see their teacher, for the added impression would aid to comprehension. Teachers were urged to be concise in speech and to present their material without ambiguities, which mislead students. The Talmud recommends group study, which allows for discussion, out of which comes greater clarity and a firmer grasp of the material studied. Humility was regarded as a prerequisite to a growth in knowledge, while arrogance was branded as its deadliest enemy.[10]

AIDS TO MEMORY

The rabbis were conscious of the danger of forgetting what had been learned at great effort. Written reference works were not plentifully available in the age before printing. They therefore created a system of mnemonic devices as an aid to memory.

A common memory aid was a well-known quotation from the Bible or some other classical text. Thus the Mishnah enumerated the feasts of Roman paganism not in their 'seasonal order, as might have been expected. It mentions them in the reverse order, the later feast being cited earlier. The verse in Ps. 139:5 "Thou hast set me behind and before" is suggested as a mnemonic for this procedure: what should have been "behind" is listed "before".

A frequently used mnemonic is a word formed from the initial letters of crucial terms that figure in the theme to be remembered. Thus the Talmud, in describing the preparation of the High Priest for the solemn Day of Atonement service, at which he was to officiate, adds that he was to confine himself to a special diet for seven days. A mnemonic is suggested to help us remember the foods which were to be avoided. These foods were citron (athrog), eggs (bezim), and old wine (yayin yashan). The initial letters in the Hebrew words denoting these foods were joined, forming the word ABY. According to another opinion his diet was also to exclude fat meat (basar shamen). By the same process of joining initial letters, and now including the word for fat meat (basar shamen), the word ABBY was formed. By the simple device of remembering ABY and ABBY we are given a clue to a readier recollection of the High Priest's diet. To cite the Talmudic text: "Symachus said in the name of R. Mari: One does not feed him either Aby, and some say, neither Abby... Aby, i.e., Athrog (ethrog, citron), nor Bezim (eggs), nor Yayin yashan (old wine). And according to others no Abby, i.e., neither Athrog (ethrog), nor Bezim, nor Basar shamen (fat meat), nor Yayin yashan."

The mnemonic occasionally consists of a key word taken from the passage that is to be fixed in memory. A good illustration of this appears in the following passage:— "(Mnemonic: Hear, And Two, Seven, Songs, Another). There was a man who used to say: Happy is a man who *hears* abuse of himself and ignores it, for a hundred evils pass him by. ... Again there was a man who used to say: Do not be surprised if a thief goes unhanged for *two* or three thefts; he will be caught in the end. ... Another used to say: *Seven* pits lie open for the good man (but he escaped); for the evil-doers there is only one, into which he falls. ... Yet another used to say: Let him who comes from a court that has taken from him his (ill-begotten) cloak sing his *song* (of relief) and go his way. ... *Another* used to say: When love was strong, we could have made our bed on a sword-blade; now that our love has grown weak, a bed of sixty cubits is not large enough for us. ..." The words listed in parentheses as the mnemonic are taken from each of the aphorisms in the passage. The word was to be a key to recall the text of the aphorism.[11]

PARABLES AND PROVERBS

The rabbis utilized parables to illustrate more vividly certain truths that they were eager to convey to their people. Scattered throughout rabbinic literature, these illustrations deal with a multitude of diverse themes. They clothe abstract ideas with concreteness, bringing them within greater comprehension by the human mind.

The masters of parable found many suggestions for their labors in the metaphors of the Bible. God is often spoken of in the Bible as King. He is king of the universe and more specifically, of Israel. This suggested many parables which explain God's ways with His creatures by reference to a king's relationship with his subjects. Israel is characterized as the Lord's first-born, and this is further clarified by stories of a king who had a dearly beloved son. The Biblical allusions to Israel as the bride of God upon whom He lavishes His love and who on occasions proves faithless to Him, inspired a series of parables about the relations of a king and the woman of his love. The story of the prophet Jonah's flight from God was further clarified by the story of the servant who sought to flee his master.[12]

The parables cited in the Talmud are for the most part centered in the moralistic sections of the literature. They are relatively absent in the discussions of law. By its very nature, the parable directs itself to the popular mind, which it seeks to impress by its homespun wisdom, rather than by formal analysis. Law was the field of interest of the scholarly community. The moralistic portions of the Talmud, on the other hand, spoke more directly to the common people.

Parables were occasionally employed in the current polemics of the rabbis against paganism. Thus Rabban Gamaliel had been asked why God's wrath is always spoken of as directed against idolators, rather than the idols. He replied by means of a parable: "A king had a son, who possessed a dog that he named after his royal father; and whenever he was about to take an oath he used to say 'By the life of the dog, the father.' When the king heard of it, at whom did he feel indignant? Against the dog or against his son? Surely against the son."[13]

Some Talmudic illustrations are fables in which animals, and occasionally plants act and speak like human beings, their experiences serving as an allegory for human life. Thus the experience of the fox in the vineyard is made to suggest the well-known truth that earthly possessions are ultimately futile since we cannot take them with us when we pass to the great beyond. The Babylonian teacher, Geniba, developed this fable in a comment in Ecclesiastes 5:14: "As he came forth of his mother's womb, naked, shall he return; as he came, so shall he go." On this Geniba commented: "This might be compared to a fox who found a vineyard which was fenced round on all sides, but it had one small hole in it. He sought to enter but he could not. What did he do? He fasted three days until he became thin and emaciated. Then he entered through the hole, and he ate and grew sleek. When he wished to leave, he could not get through that hole. He then fasted another three days until he again grew thin and emaciated and reduced to his former state, and then he went forth. On leaving he turned and gazed at the place, saying: 'O vineyard, vineyard, how goodly art thou, and how goodly is the fruit which thou producest; all thy produce is beautiful and praiseworthy, but what enjoyment have I had from thee? In the state in which one enters thee, one must leave thee'. Even so it is with the world."

Rabbi Meir is said to have employed three hundred fables in which the fox is offered as the instructor of wisdom. Only three of these have remained. Some of the fables of the Talmud show marked similarity to the fables of Aesop and the Indian moralist Kybises, but many are without parallel in other literatures.[14]

Some Talmudic illustrations are taken directly from human experience. Situations are projected in which the lesson to be taught seemed pointedly obvious, leading to its readier acceptance in the case dealt with by the rabbis.

The need for constant readiness to meet one's Maker is elaborated in a striking parable by Rabban Johanan ben Zaccai: "A king once invited his servants to a banquet without indicating the precise time when it would be given. Those who were wise remembered that things are always ready in a king's palace, and they arrayed themselves and sat by the palace gate attentive for the call to enter, while those who were foolish continued their customary occupations, saying: 'A banquet requires great preparation.' When the king suddenly called his servants to the banquet, those who were wise appeared in clean raiment and well adorned, while those who were foolish entered in soiled and ordinary garments. The king took pleasure at the wise, but was full of anger at those who were foolish, saying that those who had come prepared for the banquet should sit down and eat and drink, but those who had not properly arrayed themselves should remain standing and look on."[15]

The Talmud cites a parable which was employed by Rabbi Zera in a funeral oration, to answer the challenge of R. Abin's death, at the untimely age of twenty-eight: "A king had a vineyard in which he employed many laborers, one of whom demonstrated special aptitude and skill. What did the king do? He took this laborer from his work, and strolled through the garden conversing with him. When the laborers came for their wages in the evening, the skillful laborer also appeared among them and he received a full day's wages from the king. The other laborers were angry at this and protested: 'We have toiled the whole day, while this man has worked but two hours; why does the king give him the full wage, even as to us?' The king said to them: 'Why are you angry? Through his skill he has done in two hours more than you have done all day'. So it is with R. Abin ben Hiyya. In the twenty-eight years of his life he has attained more in the Torah than others attain in 100 years."[16]

The use of parable to offer consolation in bereavement is illustrated even more strikingly by the story concerning Beruria, wife of Rabbi Meir: "Their two sons died suddenly while Rabbi Meir was at the academy on a Sabbath afternoon. She put them on the bed and covered them with a sheet. In the evening Rabbi Meir returned and asked for the boys. She told him that they had gone to the academy. He protested that he had not seen them there. She gave him the cup of wine and he recited the prayers for the departure of the Sabbath. Then he asked once more: 'Where are our two sons?' She said to him: 'Perhaps they have gone out somewhere, but they will surely return soon.' Then she served him food and he ate. After he had eaten, she said to him: 'My master, I have a question to ask.'

He said to her: 'What is your question?' She said to him: 'O my master, the other day someone came and left in my charge a treasure, but now he has come to claim it. Shall I return it or not?' He said to her: 'Is there any question about the duty of returning property left in safekeeping to its owner?' She said to him: 'I did not want to return it without your knowing it.' Then she took him by the hand and led him to the room where the boys lay, and she placed him before the bed. She removed the sheet and he beheld the two boys lying dead on the bed. He began to cry. ... Then she told him: 'Did you not tell me that we must return the treasure to its owner?' So it is. 'The Lord hath given and the Lord hath taken, may the name of the Lord be blessed forever.' Said Rabbi Hanina: By means of that parable she comforted him and his mind became resigned to his sorrow."[17]

The rabbis found an important source of illustrations in the phenomena of nature, where they often found parallels to the phenomena of human life. The man of learning but without the necessary complement of character is compared by the rabbis to a tree laden with many heavy branches but insufficiently rooted in the earth; it lacks the sturdiness to withstand the storms ravaging the world. This illustration is quoted in the Ethics of the Fathers in the name of Rabbi Elazar ben Azarish: "He whose wisdom exceeds his works, to what may he be compared? To a tree whose branches are many, but whose roots are few; and the wind comes and plucks it up and overturns it upon its face. ... But he whose work exceeds his wisdom, to what may he be compared? To a tree whose branches are few, but whose roots are many, so that even if all the winds in the world come and blow upon it, it cannot be stirred from its place, as it is said, 'And he shall be as a tree planted by the waters; and that spreadeth out its roots by the river, and shall not perceive when heat cometh, but his leaf shall be green; and he shall not be troubled in the year of drought, neither shall he cease from yielding fruit.'" (Jer. 17:8).

The illustrations of the Talmud are often directed to the explanation of a Biblical text even as they seek to reinforce independent comments of the rabbis. Thus in accounting for the divine command which directed Abraham to leave his kin and his native land to proceed on the fateful journey to Canaan one rabbi cites the illustration of a flask of perfume: "As a flask of perfume that is hidden away in a corner gives forth no fragrance but must be poured forth to yield its fragrance, so was Abraham at the time when the Lord commanded him 'Go thee out of thy land and out of thy kindred' (Gen. 12:1). 'Abraham, Abraham', God exhorted him, 'you are a person of many noble deeds and commandments. Wander about in the world and your name will become exalted in my world.' Thus what does the verse say after the directive to set out on the journey? 'And I shall make of thee a great nation" (Gen. 12:2).[18]

Another source for Talmudic illustrations were proverbs, often drawn from popular culture. In concise and pithy formulations, often ironic in tone and peppered with humor, proverbs are copiously represented throughout Talmudic literature, and they drive home their points with a finality that no formal argument could possibly attain.

We cite here some Talmudic proverbs. Their meaning is generally self-evident, and there is no need to elucidate them by a commentary. "A person prefers one measure of his own to nine measures of his neighbor"; "Tell part of a person's praise in his presence and all of it in his absence"; "Heed your physician and you will not need him"; "The walls have ears, the woods have ears"; "Words follow the promptings of the heart"; "If the sword then not the book, if the book then not the sword"; "Who is a hero? He who can curb his passions"; "Who is wise? He who learns from all men"; "Don't consider the vessel, but what is in it"; "If your wife is short in stature, bend down and whisper to her"; "When the shepherd strays, the sheep stray after him"; "Much happens through childishness, much happens through wine"; "If Tobias sinned, shall Sigood be punished?"; "Silence becomes the wise, and surely the foolish".[19]

As a rule, the proverb is not stated independently. It is offered as additional support of some lesson that has been expounded. The following citations illustrate this: "Moses and Aaron once walked along, with Nadab and Abihu behind them, and all Israel following in the rear. Then Nadab said to Abihu, 'O that these old might die, so that you and I might become the leaders of our generation!' But the Holy One blessed be He said unto them, 'We shall see who will bury whom.' R. Papa said: Thus men say: 'Many an old camel is laden with the hides of the younger ones.'" The alleged conversation of Nadab and Abihu is a rabbinic suggestion as to what Scripture might have meant by the statement that those two had merited death because they had offered "strange fire before the Lord." (Lev. 10:1).

The identical procedure is involved in the following citation: "*The vision of Obadiah. Thus said the Lord concerning Edom* (Obadiah 1:1). Why particularly Obadiah against Edom? ... Ephraim Makshaah, the disciple of Rabbi Meir, said on Rabbi Meir's authority that Obadiah was an Edomite proselyte; and thus people say, 'From the very forest itself comes the handle of the axe that fells it.'"[20]

The rabbis did not see themselves as pioneers in the use of parable and proverb. Both appear in the Bible, principally in the writings which have been ascribed to King Solomon. They therefore commended Solomon for his contributions to this important phase of tradition. Solomon, they said, was the perfect teacher in that by means of parables, he adapted his truth to the understanding of those whom he taught. The parable, the rabbis generalized, is to abstract truth what a thread is for a labyrinth, or a trail in a thick and dark forest, or a handle to a cask of fruit or to a demijohn of boiling water, or a rope and bucket to a deep well of fresh, cold water. "Disdain it not, the parable", they added. "Remember that when a pearl of great worth is lost, we search after it with a candle that costs but the smallest coin. So the lowly parable takes us home to the great teachings of the Torah."[21]

In the style of their utterance no less than in the doctrine which they proclaimed, the rabbis regarded themselves not as innovators, but as expositors of the Scriptural word. Thus the line of development between Bible and Talmud runs clear and unbroken. In itself a vast body of literature, the Bible was also the seed for a new process of growth. And the Talmud has remained its most impressive consummation.

NOTES

INTRODUCTION

1. *Beer ha-Golah*, ch. 7.

2. Vol. V, p. 773.

3. See A. B. Tager, *The Decay of Czarism. The Beiliss Trial, Based on Unpublished Materials in The Russian Archives*, Philadelphia, 1935.

4. This work is available in the original Polish and in a German translation by Minna Safier, published in Vienna, 1937.

5. *A Commentary on the Palestinian Talmud*, N. Y. 1941, Vol. 1, pp. XXXIII, XLIII.

THE TALMUD AS LITERATURE

1. Jebamot 90b; Moed Katan 9a. ·

2. Yerushalmi Sanhedrin 4:2. ··

3. Rosh ha-Shanah 25b.

4. Baba Mezia 59b.

5. Sanhedrin 27b.

6. Mishnah, Baba Mezia 9:13; Kiddushin 24a.

7. Mishnah, Shebiit 10:2, 3. Cf. Gittin 37b.

8. Makkot 4b; Sanhedrin 54b.

9. Yoma 85b; Mishnah, Berakot 9:5.

10. Genesis Rabbah 39:21; Sanhedrin 93b; Baba Batra 15a.

11. Sanhedrin 39a; Baba Kamma 38a.

12. Mekilta on Exodus 14:15, 18:12, 21:14; Sifre on Deut. 11:13. Cf. however the analysis of S. Zeitlin, "The Halaka: Introduction to Tannaitic Jurisprudence", *The Jewish Quarterly Review*, XXXIV, I, (July 1948), pp. 14–21.

13. Mishnah, Yoma 1:1, 2, and Yoma 18a; Mishnah Yoma 8:5 and Yoma 83a.

14. Mishnah, Yoma 8:7, and Yoma 85a, 85b.

15. Mishnah, Baba Mezia 2:4, and Yerushalmi Baba Mezia 2:5.

THE FORERUNNERS OF THE TALMUD

1. Ezra 7:11, Psalm 9:2, 14, 19:2, 50:16.

2. Genesis 18:22 and Genesis Rabbah, *ad locum*.

3. Abot 1:1, 2.

4. *Perushim* is a passive construction of the verb *parosh*, but Talmudic Hebrew frequently uses the passive form with an active meaning as in Mishnah, Berakot 4:5; Tosefta Berakot 2:6, 3:18; Ketubot 26a, 66b.

5. Abot 1:4, 1:6, 1:7.

6. *Megillat Taanit* 4.

7. Josephus, *Antiquities* XIII 10:6.

8. Josephus, *Antiquities* XIII 10:5–6; Kiddushin 66a; Sanhedrin 19a.

9. Shabbat 31a.

10. Gittin 36b

11. Abot 1:10; Mishnah, Nedarim 3:4.

12. Yoma 39a.

13. Antiquities XX 9:7.

14. Mishnah, Yoma 1:1.

15. Pesahim 57a.

16. Pesahim 57a; Yoma 35b.

17. *Antiquities* XX 8:8, 9:2.

18. Josephus, *Wars* II 8:1 and *Antiquities* XVIII 1:6.

19. Josephus, *Wars* II 17:6, 16:2, 17:4.

20. Josephus, *Wars* IV 3:10, 6:1; *Vita* 35.

21. Josephus, *Wars*, Preface 2.

22. Yerushalmi Nedarim 5, end; Berakot 17a, 28b; Abot 2:8; Sukkah 28a; Tosefta Parah 4:7.

23. Gittin 56b; Abot de Rabbi Nathan, ch. 4; Rosh ha-Shanah 29b, 30a.

THE TALMUD IN ITS HISTORICAL SETTING

1. Tosefta Baba Kamma 9:30; Shabbat 151a.

2. Jebamot 63b; Tosefta Sotah 15:11–13; Baba Batra 60b.

3. Yoma 9b; Shabbat 119b, 120a; Gittin 57a.

4. Mishnah, Jebamot 1:4.

5. Baba Mezia 59a, 59b.

6. Mishnah, Abot 2:10.

7. Sanhedrin 68a, and cf. B. Z. Bokser, *Pharisaic Judaism in Transition*, N. Y. 1935, pp. 28–35.

8. Mishnah, Rosh Hashanah 2:7. The Talmud speaks of a third case in which the patriarch humiliated Rabbi Joshua.

9. Berakot 27b, 28a. See L. Ginzberg, *A Commentary on the Palestinian Talmud*, III, pp. 195f.

10. Abodah Zarah 18a.

11. Sanhedrin 74a.

12. Berakot 61b.

13. Berakot 61b.

14. Sanhedrin 14a.

15. Shir Hashirim Rabbah 2:5.

16. Ketubot 49b, 50a.

17. Kiddushin 31a; Mishnah, Gittin 5:6; Megillah 5:2; Yerushalmi, Megillah 1:4; Abodah Zarah 37a; Yerushalmi Shabbat 6:5.

18. Yerushalmi Terumot 2:3, Kiddushin 1:2, Demai 2:3, Peah 2:6, Baba Kamma 8:4; Shabbat 127a; Abodah Zarah 26b; Megillah 16a; Ketubot 111a; Genesis Rabbah 76:5.

19. Yerushalmi Nedarim 6, end.

20. Berakot 33b; Hullin 110a, 137b; Erubin 50b; Sanhedrin 5a, 64b; Yerushalmi Hagigah 1:5; Yoma 20b; Yerushalmi Rosh ha-Shanah 57a; Pesahim 50b, 113a; Yerushalmi Kiddushin 12b, end; Kiddushin 39a, 41a; Baba Mezia 59a; Baba Batra 21a, 54b; Moed Katan 24a; Genesis Rabbah 44:1

21. Shabbat 133b; Baba Mezia 85b, 108a; Baba Kamma 113b; Baba Batra 55a.

22. Baba Mezia 86a.

THEOLOGICAL ELEMENTS IN THE TALMUD

1. Genesis Rabbah 39:1

2. Hagigah 13a; Yerushalmi Hagigah 2:1.

3. Baba Mezia 31b; Sukkah 5a.

4. Genesis Rabbah 4:4.

5. Hullin 59b; Midrash Tehillim, Ps. 103:1.

6. Exodus Rabbah 5:14.

7. Genesis Rabbah 10:7; Yalkut on Proverbs, 16:4.

8. Genesis Rabbah 1:3, 12:1; Sifre Deut. section 307.

9. Genesis Rabbah 5:5; Nedarim 41a; Pesahim 118a.

10. Sifra on Leviticus 19:18.

11. Makkot 24a.

12. Tosefta Shebout 3, end.

13. Shabbat 31a.

14. Genesis Rabbah 9:7.

15. Sifre on Deut. 6:5.

16. Megillah 14a; Baba Batra 15b.

17. Baba Batra 14b, 15a; Leviticus Rabbah 15:2. The two verses alluded to are Isaiah 8:19, 20.

18. Sifre, Deut. section 343; Numbers Rabbah 13:10; Mekilta on Exodus 19:2; Sifre, Deut. section 40.

19. Sanhedrin 56a.

20. Baba Kamma 38a; Abodah Zarah 3a; Yalkut on Judges 4:1.

21. Abot 4:1.

22. Abot 2:6, 1:15, 2:9, 4:11, 6:1, 11.

23. Menahot 43b.

24. Genesis Rabbah 44:1.

25. Berakot 33b.

26. Midrash on Tehillim, Shohar Tob, on Ps. 23:5; Taanit 8a.

27. Berakot 5a, 5b.

28. Exodus Rabbah 1:1, on Ex. 27:20.

29. Sanhedrin 98b, 97a; Megillah 11a.

30. Sanhedrin 98a.

SOCIAL ETHICS IN THE TALMUD

1. Tosefta Sanhedrin 8:4.

2. Erubin 19a; Sanhedrin 101a.

3. Sanhedrin 38a; Berakot 17a.

4. Mishnah Sanhedrin 4:5; Shabbat 127a.

5. Yalkut on Judges 4:1; Gittin 61a.

6. Tanhuma on Deuteronomy 29:9; Baba Kamma 116b; Kiddushin 20a, 22b.

7. Niddah 47a; Yerushalmi Baba Kamma 6c; Mishnah, Yadaim 4:7.

8. Nedarim 49b; Abot de Rabbi Nathan II, ch. 21; Pesahim 113a; Baba Batra 110a; Kiddushin 29a; Abot 4:4; Yerushalmi Hagigah 77b; Nedarim 24b; Yoma 35b; Yerushalmi Shabbat 35a.

9. Shebuot 31a.

10. Tosefta Sanhedrin 8:3; Sanhedrin 43a, 45a; Mishnah, Makkot 1:10; Mishnah, Sanhedrin 6:5; Sanhedrin 46b.

11. Derek Erez Zuta I: Abot de Rabbi Nathan ch. 11; Sanhedrin 91b.

12. Abodah Zarah 36a; Shabbat 88a.

13. Mishnah, Eduyot 1:5; Erubin 13b; Megillah 27a; Berakot 55a.

14. Jebamot 89b; Baba Batra 8b; Sanhedrin 17b.

15. Tanhuma, Shemot, ed. Buber, p. 43a; Tosefta Peah 4:8–13; Baba Batra 8a, 9b.

16. Antiquities XX, 9:7.

17. Jebamot 62b, 63a; Kiddushin 41a.

18. Jebamot 62b; Kiddushin 30b.

19. Gittin 90b; Mishnah, Ketubot 5:6, 7:9, 10, Ketubot 77a; Mishnah, Nedarim 11:12; Mishnah, Arakin 5:6.

20. Jebamot 63b; Kiddushin 29a ff; Tosefta Niddah 2:6; Jebamot 12b.

21. Shabbat 10b; Genesis Rabbah 1; Gittin 6b; Semahot 2:6; Yerushalmi Peah 1:1.

22. Yoma 35b; Berakot 28a; Pesahim 72b.

23. Ketubot 105a; Gittin 38b; Yerushalmi Sotah 1:4; Yerushalmi Megillah 4:1.

24. Megillah 9b; Ekah Rabba 2:17, ed. Wilna, 1897; Shabbat 93b; Genesis Rabba 9, on 1:31; Sukkah 55b; Mishnah, Nedarim 3:4; Sanhedrin 39b; Tanhuma Mishpatim 12.

25. Abot de Rabbi Nathan II ch. 21, p. 22b, ed. Shechter; Sanhedrin 91b; Taanit 11a; Mekilta on Exodus 19:2; *Judaism in the First Centuries of the Christian Era*, Cambridge, 1927, I, p.324.

PERSONAL MORTALITY IN THE TALMUD

1. Berakot 60b.

2. Shabbat 32a.

3. Sotah 48b.

4. Yalkut, Beshalah, 16.

5. Abot 4:1.

6. Abot 2:16, 4:28.

7. Yerushalmi Berakot 7d.

8. Abodah Zarah 20b.

9. Sotah 4b.

10. Erubin 13b.

11. Abot 4:20, 1:15, 4:1, 2:15, 3:10.

12. Taanit 20b.

13. Yerushalmi Nedarim 9:4.

14. Genesis Rabbah, ch. 54, section 3.

15. Abot de Rabbi Nathan, ch. 23.

16. Berakot 10a.

17. Shabbat 31a.

18. Abot 1:18.

19. Numbers Rabbah 11:7; Sanhedrin 7a, 102b.

20. Abot 1:6; Abot de Rabbi Nathan ch. 41.

21. Taanit 22a.

22. Sotah 14a; Sifre, Deut. section 49.

23. Shabbat 127a; Ketubot 50a; Nedarim 39b; Megillah 3b; Berakot 18a.

24. Sukkah 49b.

25. Baba Batra 10a.

26. Leviticus Rabbah, ch. 34, end.

27. Sanhedrin 24a, 92a; Tosefta Baba Kamma 7:8; Hullin 94a; Pesahim 113h, Abot 1:18.

28. Sukkah 46b.

29. Deuteronomy Rabbah 3:5.

30. Berakot 57b.

31. Taanit 11a, 11b.

32. Berakot 29b; Numbers Rabbah 10:8.

33. Tanhuma, Noah, 14

34. Abot 2:2; Shabbat 33b.

35. Taanit 27b; Derekh Eretz 9; Shabbat 108b.

36. Leviticus Rabbah 34:3.

37. Erubin 65a; Berakot 62b; Abot 3:14; Gittin 70a; Sanhedrin 17b; Yerushalmi Kiddushin 66d.

THE JURISPRUDENCE OF THE TALMUD

1. Mishnah, Berakot 1:1, 4:3, 6:1; Berakot 28b.

2. Berakot 35a.

3. Abot 1:1.

4. Mishnah, Abodah Zarah 3, 4, 5.

5. Mekilta on Exodus 16:29; Beza 36b; Mishnah Erubin ch. 4, 5; Tosefta Shabbat, ch. 16, end (ed. Zuckermandel); Shabbat 19a.

6. Abot 5:10.

7. Abot, *ibid.*

8. Mekilta on Exodus 18:20.

9. Baba Mezia 83a; Ketubot 61a, 97a.

10. Baba Mezia 88a.

11. Baba Batra 21a, 8b; Ketubot 49b.

12. Ketubot 65b; Yerushalmi Kiddushin 1:7.

13. Baba Batra 12b.

14. Sanhedrin 40a.

15. Mishnah, Makkot 1:10.

16. Mishnah, Shabbat 6:4.

17. Mishnah, Berakot 1:1, 2, and cf. B. Z. Bokser, *Pharisaic Judaism in Transition*, N. Y. 1935, pp. 97 f.

18. Baba Kamma 113a.

19. Tanhuma, Noah, section 10.

20. Ketubot 19a; Sanhedrin 74a.

21. Sanhedrin 106b.

22. Abodah Zarah 17a.

23. Berakot 34b.

24. Mishnah, Yoma 8:9.

25. Gittin 55a.

26. Mishnah, Keritot 3:10; Baba Kamma 41b; Mishnah, Sanhedrin 9:2; Ketubot 33a; Mishnah, Peah 1:1; Yerushalmi, Pesahim 4:3; Pesahim 50a, 53a, 53b, 54b, 55a; Baba Batra 93b; Sanhedrin 23a, 23b; Soferim 14:18; Yerushalmi Baba Mezia 7:1. Rabbi Eliezer also disagreed on the number of sin-offerings required. He demanded one for each act by which the Sabbath was violated. Cf. Bokser, *Pharisaic Judaism*, pp. 129–144.

27. Niddah 61b.

28. Tanhuma, Buber, Shemini, p. 30 and cf. also Genesis Rabbah 44:1.

29. Abodah Zarah 9a.

HUMAN WISDOM IN THE TALMUD

1. Berakot 55b, 56a; Sanhedrin 30a.

2. Yerushalmi, Terumot 8, end; Ecclesiastes Rabbah 1:38; Abot 4:1, 2; Yoma 86b; Sifre on Deut. 19:11; Abot de R. Nathan 2:2.

3. Erubin 65b; Kohelet Rabbah 1; Shabbat 32a.

4. Baba Batra 29a, 31a; Jebamot 117b; Shabbat 119a; Baba Batra 5a, 5b.

5. Ketubot 13b, 59b; Megillah 13a; Moed Katan 9b.

6. Niddah 45b; Megillah 14b; Yoma 83b; Sanhedrin 67a; Genesis Rabbah 45:5.

7. Genesis Rabbah 17:7.

8. Genesis Rabbah 18:2.

9. Abot 5:12, 5:18; Leviticus Rabbah 15:2.

10. Abodah Zarah 19a; Sotah 47a; Erubin 54a, 54b, 55a; Sifre on Numbers 19:2; Erubin 13b; Abot 4:13; Pesahim 112a; Hullin 63b; Taanit 7a.

11. Abodah Zarah 8a; Yoma 7a, 18a; Erubin 54b.

12. Psalms 10:16, Zeph. 3:16, Zech. 14:16–17, Mal. 1:14 and Numbers Rabbah 2:24; Ex. 4:22, Deut. 14:1 and Berakot 13a, Deut. Rabbah 3:12, Exodus Rabbah 19:18; Isa. 54:5, Jer. 2:2, Hosea 2:18, 21, 22 and Numbers Rabbah 2:14, 15, Deut. Rabbah 3:9, 11, 16; Jonah 1 and Mekilta, Bo, ed. Weiss 1b.

13. Abodah Zarah 54b.

14. Kohelet Rabbah on 5:14; Sanhedrin 38b. Cf. Joseph L. Jacobs, *Jewish Encyclopedia*, s. v. Aesop.

15. Shabbat 153a.

16. Yerushalmi Berakot 2:8.

17. Yalkut Shimoni on Proverbs section 964.

18. Abot 3:22; Shir ha-Shirim Rabbah on 3:22. Cf. A. Feldman, *The Parables and Similes of the Rabbis*, Cambridge 1924.

19. Baba Mezia 38a, 59a; Erubin 18a; Yerushalmi Taanit 3:6; Leviticus Rabbah, 32:2; Berakot 9a; Abot 4:1, 27; Pirke de R. Eliezer 40; Abodah Zarah 17b; Sotah 7a; Makkot 11a; Pesahim 99a.

20. Sanhedrin 39b, 52a.

21. Shir ha-Shirim Rabbah 1:1.